Algorithms for Noise Reduction in Signals

Theory and practical examples based on statistical and convolutional analysis

Online at: https://doi.org/10.1088/978-0-7503-3591-1

Algorithms for Noise Reduction in Signals

Theory and practical examples based on statistical and convolutional analysis

Miguel Enrique Iglesias Martínez, Miguel Ángel García March, Carles Milián Enrique and Pedro Fernández de Córdoba

Instituto Universitario de Matemática Pura y Aplicada, Universitat Politècnica de València, Valencia, Spain

IOP Publishing, Bristol, UK

ISBN 978-0-7503-3591-1 (ebook)
ISBN 978-0-7503-3589-8 (print)
ISBN 978-0-7503-3592-8 (myPrint)
ISBN 978-0-7503-3590-4 (mobi)

DOI 10.1088/978-0-7503-3591-1

Multimedia content is available for this book from https://doi.org/10.1088/978-0-7503-3591-1.

Version: 20221201

IOP ebooks

British Library Cataloguing-in-Publication Data: A catalogue record for this book is available from the British Library.

Published by IOP Publishing, wholly owned by The Institute of Physics, London

IOP Publishing, No.2 The Distillery, Glassfields, Avon Street, Bristol, BS2 0GR, UK

US Office: IOP Publishing, Inc., 190 North Independence Mall West, Suite 601, Philadelphia, PA 19106, USA

To our parents

Contents

Preface

The present book is the result of a review of the most general techniques for the treatment of noise from the point of view of a system composed of one input and one output or two inputs and one output in the case of adaptive and artificial intelligence models and their foundations based on statistical analysis.

The first part introduces the concepts of signal and processing in a communication system, as well as different algorithms applied to noise reduction and recovery of phase information in contaminated signals. Subsequently, the book focuses on the treatment of noise using statistical processing based on nonparametric estimates of statistical characteristics such as cumulants, moments, and higher-order spectra, presenting several results from a practical point of view and in real situations.

Author biographies

Miguel Enrique Iglesias Martínez

Miguel Enrique Iglesias Martínez received a degree in Telecommunications and Electronics Engineering from the University of Pinar del Río (UPR) in 2008 and a Master's Degree in Digital Systems from the Technological University of Havana, Cuba, in 2011. In 2020 he received a Ph.D. degree in Mathematics from the Universitat Politècnica de València (UPV), receiving the Outstanding Dissertation Award to the best thesis in the area of science. He is currently at UPV, a postdoctoral researcher of the "Margarita Salas" Program, of the Ministry of Universities of the Government of Spain, assigned to the School of Aeronautical and Space Engineering of the University of Vigo. He is also associate researcher at the *Instituto Universitario de Matemática Pura y Aplicada* (UPV). His research interests include the signals processing, noise analysis, condition monitoring in electrical machines, as well as pattern recognition systems.

Miguel Ángel García March

Miguel Ángel García March received his Ph.D. in physics in 2008 from the Universitat Politècnica de València (UPV), receiving the Outstanding Dissertation Award. Currently, he is Distinguished Researcher "Beatriz Galindo" at IUMPA – Instituto Universitario de Matemática Pura y Aplicada of the Universitat Politècnica de València. Previously he was Research Fellow at ICFO – The Institute of Photonic Sciences, postdoctoral researcher at University of Barcelona (Spain) and University College Cork (Ireland), MEC/Fulbright fellow at Colorado School of Mines (US), and Postdoctoral Researcher at IFISC – Institute for Cross-Disciplinary Physics and Complex Systems, CSCI-UIB (Spain). His scientific interests include ultracold atoms, open classical and quantum systems, complex quantum dynamics, strongly correlated quantum systems, anomalous diffusion in complex classical environments, few-atom systems, nonlinear and singular optics, quantum simulators and sensors, quantum thermodynamics and relaxation in closed quantum systems. He is the author of more than 60 papers and two books.

Carles Milián Enrique

Carles Milián Enrique received his Ph.D. in physics in 2012 from the Universitat Politècnica de València, receiving the Outstanding Dissertation Award. Currently, he is a Lecturer at the Department of Applied Mathematics of the Universitat Politècnica de València. Previously he was Post-doctoral researcher at ICFO – The Institute of Photonic Sciences (Spain), at École Polytechnique de Paris (CPHT-Centre de Physique Théorique), and at the University of Bath (UK). His scientific interests are strongly focused in Nonlinear Physics with special emphasis in Nonlinear optics, solitons, dissipative systems, frequency combs, cold plasmas, and opto-mechanics.

Pedro Fernández de Córdoba

Pedro Fernández de Córdoba was born in Valencia in October 1965. He received the B.Sc., M.Sc., and Ph.D. degrees in Physics from the *Universitat de València* (UV), Valencia, Spain, in 1988, 1990, and 1992, respectively. He also received the Ph.D. degree in Mathematics from the *Universitat Politècnica de València* (UPV), Valencia, in 1997. His research work was performed at UV, UPV, the *Joint Institute for Nuclear Research* (Russia), the *University of Tübingen* (Germany), and the *Istituto Nazionale di Fisica Nucleare*, Torino, Italy, among others. He is currently Professor at the Department of Applied Mathematics (UPV) and researcher at the *Instituto Universitario de Matemática Pura y Aplicada* (UPV). His research interests include the area of modelling and numerical simulation of physical and engineering problems. Dr. Fernández de Córdoba is Doctor Honoris Causa from the *University of Pinar del Rio* (Cuba), Doctor Honoris Causa from *Universidad Santander* (México), member of the *Colombian Academy of Exact, Physical and Natural Sciences*, a member of the *Académie Nationale des Sciences, Arts et Lettres du Bénin*, Visiting Professor of the *University of Pinar del Río*, and Visiting Professor of the *Universidad del Magdalena* (Colombia). Furthermore, since its establishment on September 30th, 2011, he was member of the Board of the Spanish Mathematics-Industry network (www.math-in.net).

Glossary

ADC	analog-to-digital conversion
dB	decibel: $10\log_{10}$ (power ratio)
DWT	discrete wavelet transform
FFT	fast Fourier transform
FIR	finite impulse response
FPGA	field programmable gate array
Hz	unit of frequency in cycles per second
HOSA	higher-order statistical analysis
IIR	infinite impulse response
LMS	least mean square
MSE	mean square error
O.S	operating system
pdf	probability density function
RAM	random access memory
RLS	recursive least square
SNR	signal to noise ratio

Chapter 1

Introduction

Signals are ubiquitous in many fields of technology and even daily life. They appear in telecommunications, biomedicine, the aerospace sector, industry in general, electrical systems, etc. A signal is merely a physical magnitude quantified by means of a sensor, which mathematically takes the form of a vector that corresponds to the time series of the magnitude. Defined as such, these are the useful signals which can be used as input for many systems that require them in their functioning. But these useful input signals are always mixed with a second kind of signal, which is random, noisy, non-deterministic, and not necessarily related to the signals associated to the magnitude one aims to use. The first kind of signal, the useful one, has information which is required for the system to function. The second kind of signal, the noisy contaminating one, hinders the extraction of this information and distorts the useful signals to be processed. These noisy contaminating signals can arise from false reading of the information returned by the sensors, activation of false alarms, lack of communication between communication systems, among others. We note here that sometimes these noisy signals can be characterized and classified and may also contain useful information, as we will discuss later, mainly coded in their time or spatial correlations.

The aim of the present text is to show how to deal with these contaminating signals in practice, either reducing their distortion of the useful signal or using the information encoded in the noisy signal. One should note that, from a practical point of view, many of the contaminating noisy signals are impossible to eliminate. But it is possible to establish methods and techniques that aim to reduce their effect on the useful signals. Then, it is appropriate to devote effort to solving the problems that the disturbing signals produce in many systems of interest. In addition, the objective of the book is to provide a reasonable and updated bibliographic compilation, useful as an introduction to noise reduction methods using statistical analysis and, in particular, higher-order statistics.

doi:10.1088/978-0-7503-3591-1ch1

The recovery of a distorted signal during its transmission through a communication channel is of vital importance in digital telecommunications in particular. The examples shown in this book correspond to this kind of signal. Nevertheless one should keep in mind that the techniques described here can be adapted to any other field where one has to deal with contaminating signals. Numerous methods have been developed aiming precisely at the extraction of disturbing signals. These methods can be classified according to various criteria. In particular, and from the point of view of the number of signals available to perform the task in question, these methods can be classified into those that use *two or more input signals* and those that use *only* one *input signal.*

The methods that use only one input signal have in general the advantage of requiring less information for processing. This can be very useful for various applications where only the contaminated input signal is available. Importantly, all these methods assume some particularity of the signal to be cleaned, or of the noise to be reduced, or of some specific relationship between them (e.g. some peculiarity of the waveform of the signal to be cleaned, the particular statistical nature of the noise, information of the correlation that may exist between them, etc); these assumptions, when correct, allow us to reach better results. This book will focus on noise reduction techniques that use a single input signal for processing and will address their use on signals that exhibit *periodic behavior* contaminated with a stationary interfering signal; both signals are independent of each other. In fact, in real life environments such as telecommunications, bioengineering, and mechanical systems, it is not uncommon to find useful signals to be analyzed with very close to periodic behavior, contaminated with noise that can be considered as stationary. We will detail the requirements for the signal to which we can apply the techniques described here in chapter 3.

We propose the following classification of different types of single input signal methods of noise reduction. This classification groups the different methods according to their characteristics and the requirements they impose at the time of implementation:

- Traditional filtering methods.
- Adaptive noise reduction methods.
- Computational intelligence methods (neural networks).
- Wavelet-based methods.
- Statistical signal processing methods.

Noise reduction methods through *traditional filtering techniques* are based on treating the signal in the Fourier domain, which is the spectral signal. In this spectral domain one can identify the band of frequencies occupied by the useful signal to be cleaned. To clean the signal, one has to implement the reduction of the spectral magnitude of the noise falling outside the band occupied by the useful signal. The main advantage of this group is its low cost for practical implementations, because it is a method based on multiplications and accumulations. However, its main drawback is that the signal to be cleaned must be band-limited and the spectral content of the noise falling within this band cannot be reduced [1–11]. That is, the noise can fall inside the same band the useful signal occupies, making it impossible to disentangle the two with this kind of method.

Noise extraction techniques using *adaptive filters* require a reference or desired signal. This reference signal is in practice difficult to obtain. In some systems it is offered by the manufacturer of the system: for example, in certain heat engines or electrical machines the manufacturer can obtain this reference signal in controlled laboratory conditions or, in optimal conditions, in the first functioning of the machine. But often this pattern or reference signal cannot be provided. In this case, the reference can be the same signal but temporally shifted. Then, adaptative filters are generally based on the criterion of minimization of the error between the desired, reference signal and the output filtered signal [12–14]. That is, the output signal is adaptatively filtered to reduce the error with respect to the reference signal. To minimize this error, the method utilizes algorithms based on the gradient or steepest descent method or recursive algorithms. As noted before, for the application of these methods to the case of a single input signal it is necessary to extract a reference signal from the contaminated sample in order to evaluate the filter adaptation mechanism. However, it is well known that the conception of these techniques, and therefore their most effective application, is based on having the signal contaminated by the noise plus an additional signal correlated with the acting noise. This basic idea in its conception is a limitation for its application when only the signal contaminated by the noise is available, since from the contaminated input signal it is necessary to extract the reference to evaluate the filter adaptation mechanism. This means that this type of adaptive algorithm cannot be applied in certain bioengineering and telecommunications environments, for example, because in certain situations it is computationally expensive to extract a noise sample well correlated with the contaminating noise [15–23].

Within the *computational intelligence* methods applied to signal noise reduction, the most widely used have been *artificial neural networks*. The application of this technique is based on learning and automatic processing inspired by the way the nervous system works, so its advantages are translated into fault tolerance, flexibility, and parallel processing. However, working with neural networks is limited by the need to use architectures that incorporate the 'time' factor when applied to signal analysis. This leads to the use of very complex structures and high computational cost. One widely known technique used to work with time series is the long short-term memory, which is a particular case of an artificial recurrent neural network [24–27]. A fundamental limitation is that it is not possible to know with certainty the relationship between the final structure of the neural network and the characteristics of the signal to be cleaned or the noise to be reduced. This leads to a number of indeterminacies, for example, the definition of whether the neural architecture used is the optimal one for the application in question [28–35]. Also, the availability and quality of the training set of data determine the accuracy of the method.

As for methods based on the *signal transform*, the *wavelet transform* stands out. This is a type of mathematical transform that represents a signal in terms of translated and dilated versions of a finite wave. Its implementation is computationally faster than other transforms and is very useful for describing events that occur non-periodically. The limitation of the methods that use the wavelet transform to reduce noise is that they require knowledge of the noise magnitude (to establish thresholds) in order not to eliminate spectral components of the useful signal to be cleaned [36, 37]. There is another limitation, and it has to do with the selection of the optimal basis function for the decomposition of the contaminated input signal: it is not possible to know with certainty the relationship between the characteristics of the signal to be cleaned or the noise to be reduced, which can lead to the fact that, for example, there is no single basis function structure for noise reduction, and it can vary depending on the statistical characteristics of the interfering noise process [38–46].

With regard to *statistical signal processing methods*, it can be argued that they allow noise reduction to be performed over the entire effective working spectral band (infinite, in the case of a continuous signal; from 0 Hz to half the sampling frequency, in the case of a sampled signal). However, its traditional application is based on the assumption of linearity and stationarity, being limited to the estimation of second-order characteristics, and does not take advantage of the benefits that, for example, higher-order statistical analysis represents in many cases, as we detail below [47–54].

Here we will discuss methods based in higher-order statistical processing. These are statistical signal processing methods, but they differ from the traditional version described in the previous paragraph. The use of higher-order statistical processing has certain advantages over the methods described above. On the one hand, the contaminated input signal can be processed without the need for additional signals, which is an advantage over adaptive methods. At the same time, higher-order statistical processing does not require the performance of adaptation and learning processes, which is an advantage over methods based on neural networks; nor does it require knowledge of the magnitude of the contaminating noise, which is an advantage over methods based on wavelets. As already mentioned, noise can be reduced over the entire frequency band, which cannot be done with filtering-based methods. The most significant peculiarity of the methods based on higher-order statistical processing (cumulant calculation) is that if the noise is of a Gaussian nature, it is completely cancelled from its theoretical foundation.

Several studies have been carried out on the subject of noise reduction based on the use of higher-order statistical processing techniques [55–64]. In a majority of these works the noise of Gaussian nature is attenuated, preserving the amplitudes of the spectral components of the useful signal to be cleaned. But this comes with the drawback that the phase of these components cannot be estimated [65–68]. Other work has focused on the estimation of the phase of the spectral components of a signal contaminated by noise [69]. However, the proposals present a very high computational cost and do not focus on the estimation of the original amplitude of the spectral components of the contaminated useful signal [70–73]. Therefore, despite its advantages, the application of higher-order statistical processing to noise reduction has by no means led to obtaining the original useful signal. In essence, the reconstruction of the original signal is and will continue to be, despite all the existing advances, a problem that is in continuous development.

Here we provide and explain in detail a solution to the problem of phase recovery in periodic signals by applying higher-order statistical analysis, adopting the use of joint techniques of higher-order statistical processing and signal convolution, which is demonstrated theoretically and with practical applications. A second theoretical proposal is also described, based on a derivative-integration process in the spectral domain, for its application to signals contaminated by stationary noise, with unknown power, and with any type of probability distribution.

This book is organized as follows. In chapter 2 we describe a sample of the most current noise reduction techniques used today, following the schematic order of those mentioned in the introductory chapter. Furthermore, the theoretical foundations of these algorithms and some practical examples are also detailed, as can be seen in appendix E. Chapter 3 illustrates some of the applications of the use of the theoretical proposal based on the combination of classical statistical analysis and high-order statistics to reduce noise and obtain the phase information in a signal. Various specificities of this proposal are detailed, such as execution time and periodicity, and a comparison with several current techniques is established to validate the operation of the proposed method. In each chapter, partial conclusions of what has been addressed are offered. Likewise, finally, general conclusions and future trends of work in the area of noise reduction in signals are detailed, and the appendices show all the theoretical foundation of each chapter as well as the codes of each illustrated example.

As supplementary material in this book, all the source code of the examples shown has been included, as well as all the functions, some of which can be found in MATLAB, but in this case it has been designed for easy reproduction in embedded systems and in hardware applications (available at https://doi.org/10.1088/978-0-7503-3591-1).

References

[1] Rani S, Kaur A and Ubhi J S 2011 Comparative study of FIR and IIR filters for the removal of baseline noises from ECG signal *Int. J. Comput. Sci. Inf. Technol.* **2** 1105–8

[2] Zych M, Hanus R, Wilk B, Petryka L and Świsulski D 2018 Comparison of noise reduction methods in radiometric correlation measurements of two-phase liquid-gas flows *Measurement* **129** 288–95

[3] Arora S, Hanmandlu M and Gupta G 2020 Filtering impulse noise in medical images using information sets *Pattern Recognit. Lett.* **139** 1–9

[4] Chowdhary S and Sarde M M 2014 Review of various optimization techniques for FIR filter design *Int. J. Adv. Res. Electr. Electron. Instrum. Eng.* **3** 6566–71

[5] Roy S C D, Kumar B and Jain S B 2001 FIR notch filter design: a review *Facta Univ. Electron. Energetics* **14** 295–327

[6] Lakshmikanth S, Natraj K R and Rekha K R 2014 Noise cancellation in speech signal processing: a review *Int. J. Adv. Res. Comput. Commun. Eng.* **3** 5175–86

[7] Muangjaroen S and Yingthawornsuk T 2012 A study of noise reduction in speech signal using FIR filtering *Proc. Int. Conf. on Advances in Electrical and Electronics Engineering (Pattaya, Thailand, 13–15 April 2012)* pp 51–4

[8] Mitra S K and Kaiser J F 1993 *Handbook for Digital Signal Processing* (New York: Wiley)
[9] Shyu K-K and Chang C-Y 2000 Modified FIR filter with phase compensation technique to feedforward active noise controller design *IEEE Trans. Ind. Electron.* **47** 444–53
[10] Comon P and Pham D T 1990 Estimating the order of a FIR filter for noise cancellation *IEEE Trans. Inf. Theory* **36** 429–34
[11] Shmaliy Y S 2012 Suboptimal FIR filtering of nonlinear models in additive white Gaussian noise *IEEE Trans. Signal Process.* **60** 5519–27
[12] Bai Y, Wang X, Jin X, Su T, Kong J and Zhang B 2020 Adaptive filtering for MEMS gyroscope with dynamic noise model *ISA Trans.* **101** 430–41
[13] Di Meo G, De Caro D, Saggese G, Napoli E, Petra N and Strollo A G M 2022 A novel module-sign low-power implementation for the DLMS adaptive filter with low steady-state error *IEEE Trans. Circuits Syst.* I **69** 297–308
[14] Kumari L V R, Sabavat A J and Sai Y P 2021 Performance evaluation of adaptive filtering algorithms for denoising the ECG signal *2nd Int. Conf. on Electronics and Sustainable Communication Systems (ICESC)* pp 1–5
[15] Jimi A J, Islam M M and Mridha M M 2013 A new approach of performance analysis of adaptive filter algorithm in noise elimination *Int. J. Adv. Res. Electr. Electron. Instrum. Eng.* **2** 4571–80
[16] Ramli R M and Abid Noor A O 2012 A review of adaptive line enhancers for noise cancellation *Aust. J. Basic Appl. Sci.* **6** 337–52
[17] Hadei S A and Lotfizad M 2010 A family of adaptive filter algorithms in noise cancellation for speech enhancement *Int. J. Comput. Electr. Eng.* **2** 307–15
[18] Alves Falcão R M 2012 Adaptive filtering algorithms for noise cancellation *Master's Thesis* Faculdade de Engenharia da Universidade do Porto
[19] Sambur M 1978 Adaptive noise canceling for speech signals *IEEE Trans. Acoust. Speech Signal Process* **26** 419–23
[20] Maroofi W, Khan T and Ganar S R 2015 A review on adaptive filtering techniques for power line interference removal from biomedical signals *Int. J. Res. Appl. Sci. Eng. Technol.* **3** 689–94
[21] Ghotkar P, Bachute M and Kharadkarkar R D 2014 Noise cancellation using adaptive filtering: a review *Int. J. Curr. Eng. Technol.* **4** 3353–8
[22] Deb A, Kar A and Chandra M 2014 A technical review on adaptive algorithms for acoustic echo cancellation *Int. Conf. on Communication and Signal Processing* pp 41–5
[23] Dewasthale M M and Kharadkar R D 2014 Acoustic noise cancellation using adaptive filters: a survey *Int. Conf. on Electronic Systems, Signal Processing and Computing Technologies* pp 12–6
[24] Hochreiter S and Schmidhuber J 1997 Long short-term memory *Neural Comput.* **9** 1735–80
[25] Zhao Y, Li Y and Wu N 2022 Coupled noise reduction in distributed acoustic sensing seismic data based on convolutional neural network *IEEE Geosci. Remote Sens. Lett.* **19** 8025605
[26] Shi D, Lam B, Ooi K, Shen X and Gan W-S 2022 Selective fixed-filter active noise control based on convolutional neural network *Signal Process.* **190** 108317
[27] Song H, Kim M, Park D, Shin Y and Lee J-G 2022 Learning from noisy labels with deep neural networks: a survey *IEEE Trans. Neural Netw. Learn. Syst.* accepted
[28] Bactor P 2012 Review of adaptive noise cancellation techniques using fuzzy-neural networks for speech enhancement *Int. J. Comput. Corp. Res.* **2** 1

[29] Kaur B and Dhir V 2013 Neural network based new algorithm for noise removal and edge detection: a survey *Int. J. Innov. Res. Sci. Eng. Technol* **2** 5209–14

[30] Bagheri F, Ghafarnia N and Bahrami F 2013 Electrocardiogram (ECG) signal modeling and noise reduction using Hopfield neural networks *Eng. Technol. Appl. Sci. Res.* **3** 345–8

[31] Badri L 2010 Development of neural networks for noise reduction *Int. Arab J. Inf. Technol.* **7** 289–94

[32] Zeng X and Martinez T 2003 A noise filtering method using neural networks *IEEE Int. Workshop on Soft Computing Techniques in Instrumentation, Measurement and Related Applications* 2003 26–31

[33] Maas A L, Le Q V, O'Neil T M, Vinyals O, Nguyen P and Ng A Y 2012 Recurrent neural networks for noise reduction in robust ASR *Proc. Interspeech* **2012** 22–5

[34] Kosko B 1990 Unsupervised learning in noise *IEEE Trans. Neural Netw.* **1** 44–57

[35] Zhang X-P 2001 Thresholding neural network for adaptive noise reduction *IEEE Trans. Neural Netw.* **12** 567–84

[36] Rhif M, Abbes A B, Farah I R, Martínez B and Sang Y 2019 Wavelet transform application for/in non-stationary time-series analysis: a review *Appl. Sci.* **9** 1345

[37] Mgaga S S, Khanyile N P and Tapamo J-R 2019 A review of wavelet transform based techniques for denoising latent fingerprint images *Open Innovations (OI)* 57–62

[38] Pizurica A, Wink A M, Vansteenkiste E, Philips W and Jos Roerdink B T M 2006 A review of wavelet denoising in MRI and ultrasound brain imaging *Curr. Med. Imaging Rev.* **2** 247–60

[39] Singh M and Garg N K 2014 Audio noise reduction using wavelet types with thresholding techniques *Res. Cell Int. J. Eng. Sci.* **3** 86–8

[40] Ghosh T, Bhattacharyya D, Bandyopadhyay S K and Kim T-H 2014 A review on different techniques to de-noise a signal *Int. J. Control. Autom.* **7** 349–58

[41] Baleanu D (ed) 2012 *Advances in Wavelet Theory and Their Applications in Engineering* (London: IntechOpen)

[42] Dewangan N and Bhonsle D 2013 Comparison of wavelet thresholding for image denoising using different shrinkage *Int. J. Emerg. Trends Technol. Comput. Sci.* **2** 57–61

[43] Cohen R 2012 *Signal Denoising Using Wavelets* Department of Electrical Engineering Technion, Israel Institute of Technology, Haifa

[44] Ouahabi A 2013 A review of wavelet denoising in medical imaging *8th Int. Workshop on Systems, Signal Processing and Their Applications (WoSSPA)* pp 19–26

[45] Chen G 2013 Wavelet-based denoising: a brief review *4th Int. Conf. on Intelligent Control and Information Processing* pp 570–4

[46] Cohen I and Berdugo B 2002 Noise estimation by minima controlled recursive averaging for robust speech enhancement *IEEE Signal Process. Lett.* **9** 12–5

[47] Shao M and Nikias C L 1993 Signal processing with fractional lower order moments: stable processes and their applications *Proc. IEEE* **81** 986–1010

[48] Sandgren N, Stoica P and Babu P 2012 On moving average parameter estimation *Proc. 20th European Signal Processing Conf. (EUSIPCO)* pp 2348–51

[49] Bowden R S and Clarke B R 2012 A single series representation of multiple independent ARMA processes *J. Time Ser. Anal.* **33** 304–11

[50] Chen Y, Xu Y, Mierop A J and Theuwissen A J P 2012 Column-parallel digital correlated multiple sampling for low-noise CMOS image sensors *IEEE Sens. J.* **12** 793–9

[51] Sharma V and Dhaka K D 2014 Review paper on second order statistics of various fading channels *Int. J. Adv. Res. Electr. Electron. Instrum. Eng* **3** 10885–7
[52] Sanaullah M 2013 A review of higher order statistics and spectra in communication systems *Glob. J. Sci. Front. Res.* **13** 31–50
[53] El Helou M and Süsstrunk S 2020 Blind universal Bayesian image denoising with Gaussian noise level learning *IEEE Trans. Image Process* **29** 4885–97
[54] Fang H, Tian N, Wang Y, Zhou M-C and Haile M A 2018 Nonlinear Bayesian estimation: from Kalman filtering to a broader horizon *IEEE/CAA J. Autom. Sin.* **5** 401–17
[55] Green D R 2003 The utility of higher-order statistics in Gaussian noise suppression *Master's Thesis* Naval Postgraduate School
[56] Cao T, Zhao X, Yang Y, Zhu C and Xu Z 2022 Adaptive recognition of bioacoustic signals in smart aquaculture engineering based on r-sigmoid and higher-order cumulants *Sensors* **22** 2277
[57] Takahashi Y, Miyazaki R, Saruwatari H and Kondo K 2012 Theoretical analysis of musical noise in nonlinear noise reduction based on higher-order statistics *Proc. 2012 Asia Pacific Signal and Information Processing Association Annual Summit and Conf.*
[58] Giannakis G B and Dandawate V A 1991 Higher-order statistics-based input/output system identification and application to noise cancellation *Circuits Syst. Signal Process.* **10** 485–511
[59] Rosenblattm M and Van Ness J W 1965 Estimation of the bispectrum *Ann. Math. Stat.* **36** 1120–36
[60] Huber P, Kleiner B, Gasser T and Dumermuth G 1971 Statistical methods for investigating phase relations in stationary stochastic processes *IEEE Trans. Audio Electroacoust* **19** 78–86
[61] Benveniste A, Goursat M and Ruget G 1980 Robust identification of a nonminimum phase system: blind adjustment of a linear equalizer in data communications *IEEE Trans. Autom. Control* **25** 385–99
[62] Raghuveer M and Nikias C 1985 Bispectrum estimation: a parametric approach *IEEE Trans. Acoust. Speech Signal Process.* **33** 1213–30
[63] Khoshnevis S A and Sankar R 2020 Applications of higher order statistics in electro-encephalography signal processing: a comprehensive survey *IEEE Rev. Biomed. Eng.* **13** 169–83
[64] Trapp A and Wolfsteiner P 2021 Estimating higher-order spectra via filtering-averaging *Mech. Syst. Signal Process.* **150** 107256
[65] Giannakis G B and Mendel J M 1989 Identification of nonminimum phase systems using higher order statistics *IEEE Trans. Acoust. Speech Signal Process.* **37** 360–77
[66] Anderson J M M, Giannakis G B and Swami A 1995 Harmonic retrieval using higher order statistics: a deterministic formulation *IEEE Trans. Signal Process.* **43** 1880–9
[67] Sharma L N, Dandapat S and Mahanta A 2013 Kurtosis-based noise estimation and multiscale energy to denoise ECG signal *Signal Image Video Process.* **7** 235–45
[68] Anjaneyulu L, Murthy N S and Sarma N V S 2009 A novel method for recognition of modulation code of LPI radar signals *Int. J. Recent Trends Eng.* **1** 176
[69] Xiang Q, Yang Y, Zhang Q, Fan Q and Yao Y 2019 Machine learning enhanced modulation format transparent carrier recovery based on high order statistics *45th European Conf. on Optical Communication (ECOC 2019)*
[70] Hsieh H-Y, Chang H-K and Ku M-L 2011 Higher-order statistics based sequential spectrum sensing for cognitive radio *11th Int. Conf. on ITS Telecommunications* pp 696–701

[71] Buades A, Coll B and Morel J-M 2005 A review of image denoising algorithms, with a new one *Multiscale Model. Simul.* **4** 490–530
[72] Sharmila V and Ashoka Reddy K 2014 Modeling of ECG signal with nonlinear Teager energy operator *Int. J. Comput. Sci. Netw. Secur.* **14** 34–41
[73] Hernández Montero F E 2000 Aplicación de redes neuronales artificiales en la cancelación de ruido *Master's Thesis* Faculty of Automation, Technological University of Havana, Cuba

Chapter 2

Current trends in signal processing techniques applied to noise reduction

Digital signal processing deals with the modeling, detection, identification, and utilization of patterns and structures of a signal [1]. Applications of signal processing methods include audio, digital radio and television, mobile telephony, voice recognition, vision, radar, geophysical exploration, and medical electronics. In general it can be used in any system where communication or information processing is involved. Signal processing theory plays a fundamental role in the development of digital telecommunication and automation systems and in the efficient transmission, reception, and decoding of information [1]. The theory of statistical signal processing provides the foundation for the description of a random process in environments in which signals propagate. Statistical models are applied in signal processing and decision-making systems for extracting information from a signal that may be noisy, distorted, or incomplete.

This chapter will provide an introduction to the concepts of signal and processing in a communication system. We will also introduce the fundamentals of higher-order statistical analysis and its application to noise reduction and recovery of phase information in contaminated signals.

2.1 Signals and noise

A *signal* can be defined as the quantitative variation of a magnitude, such as temperature of a body, current in a conductor, color of an image, etc. For this quantitude the information is carried with respect to one or more attributes of the source, such as state, characteristics, composition, and trajectory. From a different point of view, a signal is the carrier of information or the transmission medium of the information associated to the magnitude. Then, the information contained in

doi:10.1088/978-0-7503-3591-1ch2

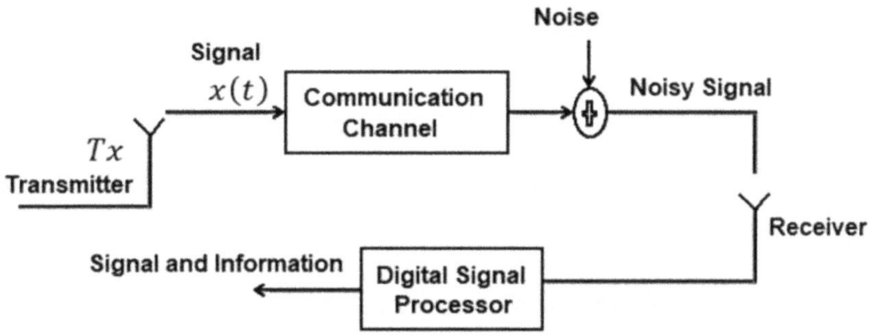

Figure 2.1. Schematic representation of an exemplary communication and signal processing system.

the signal can be used for various applications. Figure 2.1 presents a schematic of a communication and signal processing system. The magnitude information is present in the transmitter and produces the signal. Then this signal travels through a communication signal. One source of noise occurs during this communication channel, where noise is added. Then the noisy signal arrives at the receiver, and after that, this signal is processed to obtain the useful information. Let us note that it is also possible for noise to occur in the transmitter or the receiver, which is known as instrumentation noise. Often, this noise is due to malfunctioning of the instrumentation or its wrong use. For example, when a medical signal is generated, say in an electrocardiogram, the patient should not introduce unwanted noise by excessive movement etc. Other examples are unwanted vibrations or wrong calibration of the instrumentation. All these sources of noise can be cured with appropriate use of the instrumentation. For this reason here we consider mainly those noise sources which cannot be avoided, such as noise added through the communication channel.

In general, digital signal processing focuses on two broad areas [2]:

- Efficient and reliable coding, transmission, reception, storage, and representation of signals in communication systems.
- Extraction of information from noisy signals for recognition of contaminant signal characteristics, detection, decision-making, control, and automation.

Particularly the second area, referred to as the extraction of information from noisy signals, is the main objective of this book.

To be specific, let us define *noise* as an unwanted signal that interferes with the communication or measurement (transmitter or receiver, that is, the instrumentation) of another signal. A noise itself is a signal carrying information regarding the noise sources and the environment in which it propagates [2]. The typology of algorithms used for noise reduction, which was introduced in chapter 1, is discussed in detail below. We also will describe the concepts required to understand the treatment of noise using higher-order statistical analysis.

2.2 Current trends in signal processing techniques applied to noise reduction

From what was discussed in the introductory chapter 1, it can be said that a wide range of techniques using a single input signal have been applied for noise reduction in periodic signals. This is because there is no single algorithm or general method which works appropriately for noise reduction in every signal or information system. The random nature of the interfering process depends on dissimilar characteristics of both the application environment and statistics of the analyzed process which makes it impossible to find a single method that works for any kind of signal. In table 2.1 we summarize the typology of methods introduced in chapter 1 and include the related references for the interested reader. Each group of techniques and their performance in application to noise reduction are presented in the following sections, and the advantages and disadvantages of each particular group are discussed. We only discuss the digital processing of signals, that is, of discrete signals. Then, we particularize the traditional filtering methods as those based on finite impulse response (FIR) and infinite impulse response (IIR) system impulse response.

2.2.1 Filtering methods based on FIR and IIR system impulse response

The term *filter* is commonly used to describe a device that discriminates what passes through it [19]. A filter is a linear and time invariant system (linear time invariant), which modifies the frequency spectrum of the *input signal* according to its frequency response $H(\omega)$ (known as the transfer function), where ω is the so-called *normalized discrete frequency*. The filter produces an *output signal* which can be further processed [19].

In general, there are two ways to describe the impulse response of a linear and time invariant system: (i) FIR and (ii) IIR [19]. The FIR systems are characterized by being non-recursive, and the IIR systems are distinguished by having feedback in the output signal.

Let $x(n)$ be the input signal and $y(n)$ the output signal, with n labeling the entries in the discretized input signal and output signal vectors. A digital filter of order N can be described by the difference equation [19]

Table 2.1. Summary of the typology of methods and related literature.

Method	References
Filtering methods based on FIR and IIR system impulse response	[3–6]
Adaptive noise reduction methods	[7–9]
Computational intelligence methods—neural networks	[10–12]
Wavelet-based methods	[13–15]
Statistical signal processing methods	[16–18]

$$\sum_{m=0}^{N} a_m \cdot y(n-m) = \sum_{k=0}^{N} b_k \cdot x(n-k), \tag{2.1}$$

where a_m, b_k are the filter coefficients and N the filter order. The requirements for the synthesis of a digital filter are usually specified in the frequency domain in terms of desired amplitude and phase response [19]. For example, for the case of a low pass filter, the frequency response of the filter is given by [19]

$$H(\omega) = \begin{cases} 1, & \forall\, \omega \in \left[0,\, \omega_{\mathrm{p}}\right], \\ 0, & \forall\, \omega \in [\omega_{\mathrm{s}},\, \pi], \end{cases} \tag{2.2}$$

where $\left[0,\, \omega_{\mathrm{p}}\right]$ refers to the pass band of the filter and $[\omega_{\mathrm{s}}, \pi]$ to its rejection or stop band. The region $[\omega_{\mathrm{p}}, \omega_{\mathrm{s}}]$ is the transition band.

This technique has the advantage of an easy practical implementation since it is based on multiplications and accumulations. It has the disadvantage of the impossibility of attenuating the noise included in the frequency band of the useful signal, that is, the pass band. Figure 2.2 shows the frequency response of a low pass filter. Where the filter pass frequency is represented on the abscissa axis and the filter amplitude response is shown on the ordinate axis, respectively, ω_{p} and ω_{p} represent the pass and stop frequency respectively. Likewise δ_{p} and δ_{s} are the pass and stop impulse response amplitude.

The way to reduce noise using this type of filter is basically based on analyzing the spectral content of the input signal according to the frequency response of the filter. That is, there may be noise in the three bands of the frequency response of the filter, but the filter can only eliminate what is outside the pass band of the filter: the incident noise inside the pass band cannot be reduced.

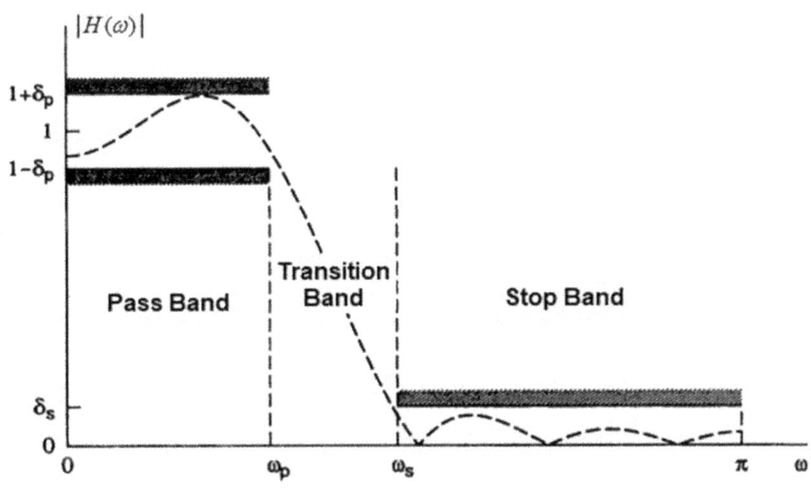

Figure 2.2. Frequency response of a low pass filter: pass band, transition band, and rejection or stop band.

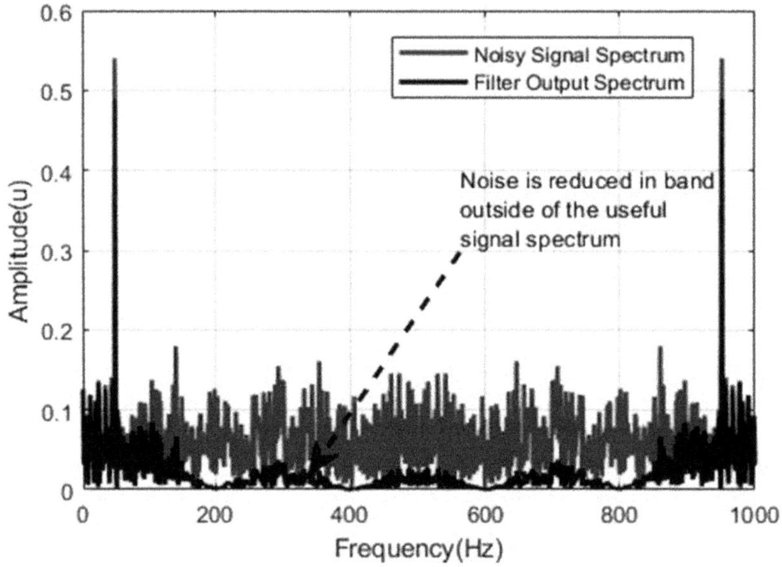

Figure 2.3. Noisy signal and filter output spectrum.

As shown in figure 2.2, if the contaminating noise has spectral content in a band that coincides totally or partially with the pass band of the filter, it cannot be reduced. To illustrate an example of the above, we have the following. Let $x(t) = \sum_{k=1}^{N} A_k \cos(w_k t + \phi_k)$ be a stationary real signal of zero mean value constituted by a sum of N harmonics, of which $w_c = 2\pi f_c$ is the principal component signal frequency, in this case 50 Hz; ϕ_k is the phase, a statistically independent and uniformly distributed random variable of $0 \leqslant \phi \leqslant 2\pi$ with pdf $= \frac{1}{2\pi}$; and A_k is the amplitude of each harmonic. Then for $N = 1$, and adding noise to the signal, the described process can be written as follows:

$$y(t) = x(t) + n(t), \tag{2.3}$$

where $n(t)$ is the additive noise. Figure 2.3 shows the results of applying a moving average filter to the model described in equation (2.3). As can be seen, only the noise outside the useful frequency band is reduced; this is not the case with the noise within the useful frequency band around the fundamental frequency. The complete code of the example and the description of the moving average algorithm used can be found in appendix E.

2.2.2 Adaptive noise reduction methods

Currently, there are applications in digital communications that have noise-related limitations which generate errors. One example is intersymbol interference, which is when the noisy signal falls within the pass band and cannot be reduced with traditional methods, that is, the filtering methods described in the previous section.

Other examples are echo noise, which is the same signal but displaced, and additive noise, which is the most common one and simply adds to the useful signal. These problems can be solved through the use of adaptive filters [20].

Adaptive filters are intimately related with the impulse response–based filtering methods discussed in the previous section. They add the particularity of incorporating an adaptation process. The main characteristic of these filters is that they can modify their response during operation in order to achieve a targeted behavior. For this, it is necessary to use of an adaptive algorithm and to have access to a reference signal or to be able to construct it from the information at hand, as we describe below. This algorithm allows us to update the filter coefficients during operation [20]. The best known adaptive algorithms are the recursive least square (RLS) algorithm, the least mean square (LMS) algorithm, and the Wiener filter (see more examples in [2]). The RLS algorithm offers higher convergence speed with respect to the LMS algorithm. On the other hand, the LMS algorithm has the advantage of being simpler in computational terms.

A great advantage of the adaptive noise reduction methods is that they can adapt their performance in the presence of dynamic environments, that is, in systems where the noise characteristics change over time.

The main disadvantage of these methods is that they require more than one input signal for processing. This basic idea in its conception is a limitation to be applied when only the signal contaminated by noise is available, since from the contaminated signal input it is necessary to extract the reference to evaluate the filter adaptation algorithm.

A single-input adaptive filter such as the one shown in figure 2.4 has four terminals: (i) the polluted input signal, $x(n)$; (ii) the reference signal, $d(n)$; (iii) the filtered output signal, $y(n)$; and (iv) the error signal, $e(n)$. The latter is used to adapt the filter coefficients, referred to as $w(n)$. In the adaptive filtering structure used to reduce noise, the output of the system corresponds to the error signal. Importantly, in these systems, the reference signal can be obtained by filtering [2], delay mechanisms, or other particular algorithms capable of extracting a signal correlated to a greater or lesser extent with the noise present in the contaminated input signal.

Figure 2.5 shows an example of adaptive filter signal processing. As can be seen, the noise is reduced in the useful signal band.

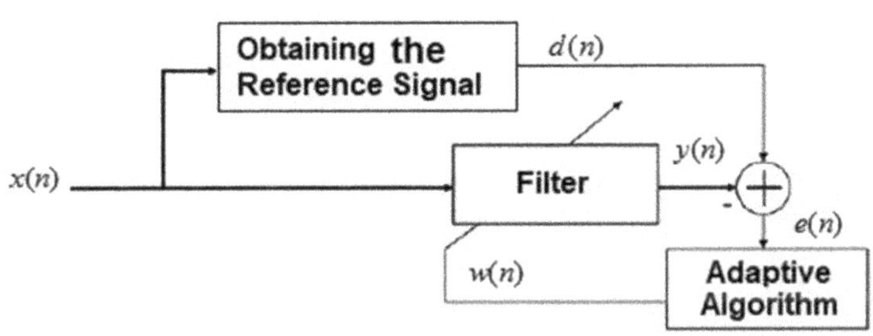

Figure 2.4. Schematic of the adaptive filtering method to reduce noise [20].

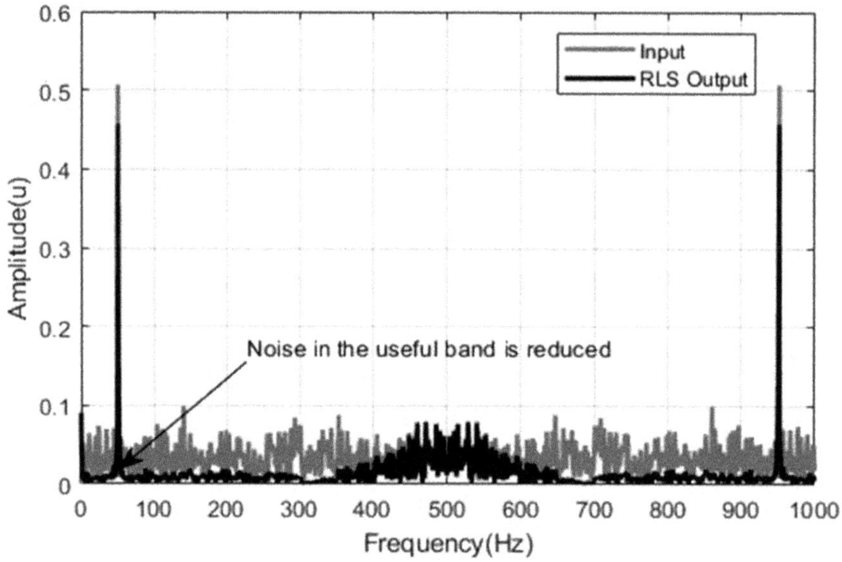

Figure 2.5. Input–output comparison spectral domain RLS.

In appendix E we provide the code that implements this (see in that appendix examples of adaptive filter implementation).

2.2.3 Machine learning methods: neural networks

Artificial neural networks, using a parallel, distributed, and adaptive computing, are capable of learning from a large set of training instances. These systems schematically mimic the hardware (neural) structure of the brain in an attempt to reproduce some of its capabilities. In practice, an artificial neural network can be simulated by a computer program or realized in specific electronic circuits [21].

The idea of using neural networks for noise reduction, in its general form, is an approach very similar to that employed by neural architectures involved in pattern recognition. An example of this essentially consists of training the chosen neural network architecture so that it learns the statistics of the random process, immersed in which the signal to be processed is found [22]. From a pattern recognition approach, the network is provided with samples of certain noise (similar to training patterns in a pattern recognition method), assuming that the expected value of that random process is obtained at the output. With these data, the network is trained several times [22, 23].

Once the architecture is trained, it is operated by introducing as input the useful signal contaminated by noise (of the same type as those which were used during training). If the network is trained correctly, the output will be data that match the average or expected value (desired signal). Figure 2.6 shows an exemplary operation of a neural network for use in noise reduction tasks [21]. Note that the delays applied to the introduced contaminated signal play a role similar to the finite difference

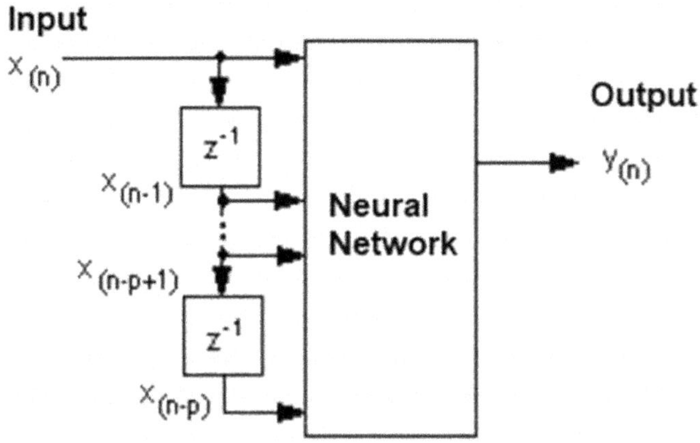

Figure 2.6. An example of a neural network used to reduce noise. Here $x(n)$ is the contaminated signal, $y(n)$ is the filtered output signal, and Z^{-1} is a delay of one time step (one entry in the vector)—note that applying p times Z^{-1} produces a delay of p steps.

equation (2.1) used in filtering methods, heuristically. Here there are no coefficients; they are introduced within the neural network, which acts here as a black box.

Regarding this group of techniques, machine learning methods present the advantages of having an adaptation and learning mechanism, fault tolerance, flexibility, and parallel processing. The disadvantage is that learning requires a training time that can be high, depending on the structure or algorithm selected. In references [24–26] one can consult codes implementing these methods. In addition, appendix E shows the implementation of a trained network and its application to noise reduction in an image.

2.2.4 Wavelet-based methods

A wavelet is a signal of limited duration whose mean value is zero. Comparing wavelets with sinusoidal functions (which are the basis of Fourier analysis), it can be highlighted that the main difference lies in the fact that sinusoidal signals do not have limited duration, since they extend from $-\infty$ to $+\infty$ [27]. Moreover, while sinusoidal signals are smooth and predictable, wavelets tend to be irregular and asymmetric [27].

Given a base function $\Psi(t)$, the continuous wavelet transform is given by the expression [27, 28]

$$X(a, b) = \frac{1}{\sqrt{a}} \int_{-\infty}^{\infty} \Psi\left(\frac{t-b}{a}\right) \cdot x(t)dt. \quad (2.4)$$

Here, the parameter a is the frequency information, the parameter b refers to the temporal information in the transform, and $x(t)$ is the input signal [29]. In practice, for a finite number of samples, the discrete wavelet transform (DWT) is used. A common approach to the denoising problem is as follows. Suppose there are n noisy samples of a function f, that is,

$$x_i = f(t_i) + \sigma \mathrm{E}_i, \tag{2.5}$$

with $i = 1, \ldots, n$ and where the E_i are statistically independent variables and the noise level $\sigma\mathrm{E}_i$ may or may not be known. Here f is the useful signal we aim to extract, and x is the contaminated signal.

We now apply the DWT to x_i. This produces the coefficients w_{jk} where j denotes the level of decomposition and k is the index of the coefficient at this level of decomposition. Since the orthogonality of the DWT implies that white noise transforms into white noise [28] one also obtains the noise in the resulting transformed signal. That is, if x_{jk}, where j denotes the level of decomposition and k is the index of the coefficient at this level of decomposition, are the wavelet coefficients of the contaminated signal x_i, the transform of equation (2.5) for the wavelet domain is [28]

$$x_{jk} = w_{jk} + \sigma E_{jk}, \tag{2.6}$$

where w_{jk} are the wavelet coefficients of the noise-free useful signal sample, $f(t_i)$, and E_{jk} are the wavelet coefficients of the observed data, which can be considered as the noisy sample.

The coefficients of the wavelet transform are usually sparse, meaning that most of the coefficients in a noiseless wavelet transform are effectively zero. The problem of obtaining the useful signal $f(t_i)$ can then be reformulated as obtaining those coefficients w_{jk} that are relatively larger in amplitude than the noise. From this wavelet coefficient one can obtain the output signal, that is, the useful signal. This means that the coefficients of small amplitude can be considered as noise and are made zero.

The method where each coefficient of the wavelet transform is compared with a threshold in order to decide whether or not it is part of the original signal is called wavelet thresholding [30].

The thresholding of the wavelet coefficients extracts the significant coefficients to the useful signal by making zero the coefficients whose absolute value is less than a certain threshold level. This threshold value can usually be a function dependent on the decomposition levels of the input signal. Most algorithms using the thresholding criterion try to estimate the optimal value of the threshold. However, the first step in these methods usually involves estimating the noise level σ, assuming that σ is proportional to the standard deviation of the coefficients, which is not a good estimator unless the spectrum f is flat. With respect other methods, the wavelet methods have the advantage of presenting a faster implementation from the computational point of view than other transforms, besides being useful to describe events that occur non-periodically. Their main disadvantage is that, to reduce the noise, it is necessary to know the power of the interfering signal in order to obtain a noise threshold at the time of processing.

An example of uses of wavelet denoising methods is shown in figure 2.7. The exemplary codes of this can be consulted in appendix E.

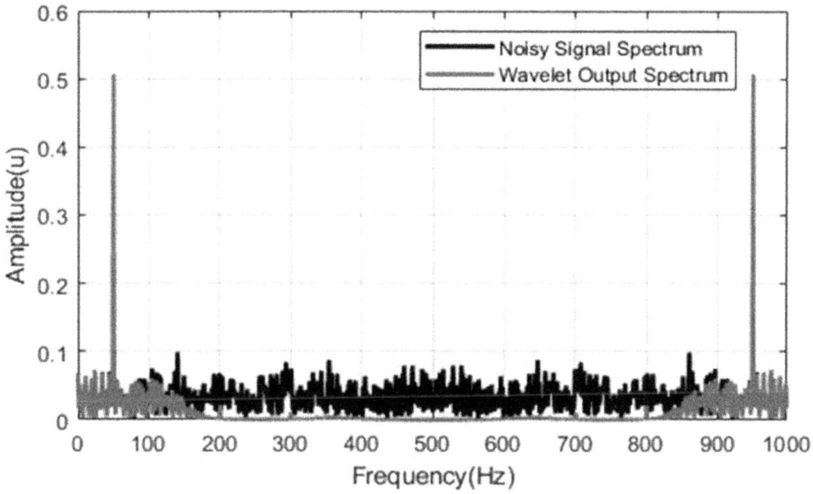

Figure 2.7. An example of a wavelet denoising methods.

2.3 Introduction to higher-order statistical analysis

Here, we introduce statistical signal processing methods, which have the following advantages: (1) they can process the input contaminated signal without a reference sample or additional information; (2) they do not require adaptation and learning processes to perform the task of reducing the noise in the signal; and (3) it is not necessary to know the magnitude of the noise, and the noise can be reduced in the whole frequency band. Moreover, the noise becomes zero if it is Gaussian in nature. This is advantageous with respect to the other noise reduction techniques, which is why it will be the algorithm described in the rest of this book.

We will use higher-order statistical analysis (HOSA) techniques with periodic signals contaminated with a stationary interfering signal. A stationary signal is characterized by constant statistics during the duration of the useful signal or at least during a period of the periodic useful signal. Being more specific, the mean, variance, etc which characterize the statistical behavior of the random noise signal are constant in time. We will treat both white and colored noise. The colors of noise are a way of denominating possible different temporal autocorrelations in the noise. White noise is a signal $W(t)$ which is originated by a random process. A random process is a physical process that originates a variable (a magnitude) which takes stochastic values according to some probability distribution. The two main groups of random processes are the Wiener processes (associated to Brownian motion) and the Poisson processes. Then, the white noise, which is the random variable $W(t)$, is a random process which fulfills the following:

1. The mean of $W(t)$ is constant and zero.
2. if $t_1 \neq t_2$ then $W(t_1)$ is independent from $W(t_2)$; that is, it in uncorrelated in time (or space). For physicists, this can be expressed as $\langle \xi(t), \xi(t') \rangle = \delta(t - t')$, with $\langle \cdot, \cdot \rangle$ denoting the scalar product. In the context

of this book, we will write the autocorrelation as the cumulant of second order, which is explicitly given as

$$c_2^W(\tau_1) = E\{W(t)W(t + \tau_1)\} \tag{2.7}$$

3. The variance of $W(t)$ is constant.
4. $W(t)$ is independent from any other variable in the system.

The white noise has constant power spectral density, which is the modulus of the Fourier transform of the autocorrelation function [31].

We will also consider colored noise, where the time autocorrelation is not zero and hence results in a non-constant power spectral density. The form of the power spectral density determines the kind of colored noise considered. For example, the frequency spectrum of *pink noise* is linear in logarithmic scale. Then there are colors of noise with a growing linear in logarithmic scale spectral density, e.g. blue, violet, etc. Non-monotonic spectral density includes grey noise, the structure of the noise associated with environmental stochasticity. White noise, uncorrelated in time, corresponds to an environment where future environmental conditions are not related with past environmental conditions. When the noise is colored, the environmental conditions are related in time and have positive correlation (red or pink noises) or negative correlation (blue noise). When we refer to Gaussian noise we mean white noise which is extracted from a Gaussian probability distribution with zero mean. When we refer to Gaussian colored noise, it is equally extracted from a Gaussian probability distribution with zero mean, but it shows non-zero autocorrelation. We emphasize that in this book we only consider additive noise, that is, $x(t) = f(t) + \xi(t)$, with $x(t)$ the signal to be treated being equal to $f(t)$, the useful signal plus a noise, $\xi(t)$, with the noise produced by some stochastic process. The noise can be integrated in the useful signal in other ways, for example, multiplicative noise or impulsive noise [32, 33].

Finally, we note that the methods described below may be used also for transient (non-periodic) signals, always assuming the (white or colored) noise is stationary, with an appropriate treatment of data in intervals of stationarity. In appendix E an example of noise reduction in radio frequency pulse is illustrated.

We also may consider a nonlinear useful signal, meaning that the signal is generated by a nonlinear system; that is, the differential equations which describe the dynamics of the system are nonlinear (if the signal is $x(t)$); the differential motion equations include powers of x. Some examples of several types of noise are illustrated in figure 2.8; the code to obtain this figure can be founded in appendix E.

Higher-order spectral analysis, also known as *polyspectra*, is defined in terms of higher-order statistics (cumulants, in particular). Specific cases of higher-order spectra include the third-order spectrum, also called the bispectrum or Fourier transform of the third-order time cumulant, and the trispectrum or Fourier transform of the fourth-order time cumulant, for a given signal. Figure 2.9 illustrates the

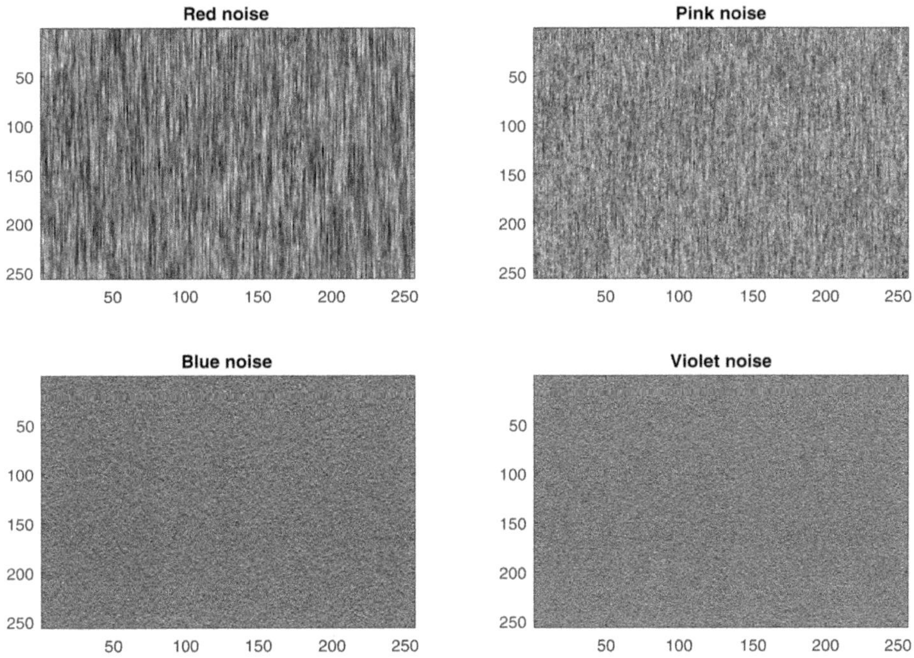

Figure 2.8. Classification of several types of noise.

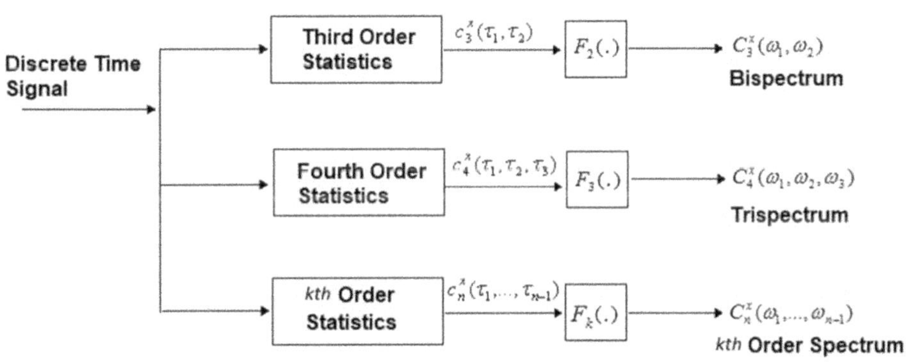

Figure 2.9. Classification of higher-order spectra of a discrete signal $x(n)$. Here, $F_k(.)$ denotes the Fourier transform of dimension k [34].

classification of the higher-order spectra of a given signal. Although the higher-order statistical characteristics and spectrum of a signal can be defined in terms of moments and cumulants, the former and their spectra can be very useful in the analysis of deterministic signals (transient and periodic useful signals), while the cumulants and their spectra are of great importance in the analysis of stochastic signals, i.e. the noise [34].

There are several general motivations behind the use of higher-order spectra in signal processing. These methods can be used to (1) remove Gaussian additive colored noise from an unknown power spectrum; (2) extract information due to deviations from processes whose probabilistic distribution function is Gaussian; and (3) detect and characterize nonlinear properties in signals, as well as to identify nonlinear systems.

The first motivation is based on the property that, for stationary signals with Gaussian probabilistic distribution function, the higher-order cumulants are equal to zero. If a non-Gaussian signal is received together with additive Gaussian noise, a calculation of the higher-order cumulant of the signal plus noise sample will eliminate the noise. Therefore, in these signal processing environments there will be certain advantages for the detection and/or estimation of signal parameters using the cumulant of the observed data [34].

2.3.1 Higher-order statistics: definition and properties

This section introduces the definitions, properties, and calculation of higher-order statistics.

Let $\{X(k)\}$, $k = 0, \pm1, \pm2, \pm3, \dots, N$ be a stationary random vector of length $N + 1$. We recall that this vector includes the useful signal and the noise. Let us assume that its higher-order moments exist. Then

$$m_n^X(\tau_1, \tau_2, \dots, \tau_{n-1}) = E\{X(k)X(k + \tau_1)\cdots X(k + \tau_{n-1})\},$$

with $E\{\cdot\}$ the expected value [35], represents the nth-order moment of that vector. It depends only on the different time spans $\tau_1, \tau_2, \dots, \tau_{n-1}$ $\tau_i = \pm j$ for all j up to $N + 1$. Clearly it can be seen that the second-order moment $m_2^X(\tau_1)$ is the autocorrelation function of $\{X(k)\}$ (see (2.7) where we defined it for the noise $W(t)$, but can be also applied for the signal). Likewise $m_3^X(\tau_1, \tau_2)$ and $m_4^X(\tau_1, \tau_2, \tau_3)$ represent the third- and fourth-order moments, respectively.

On the other hand, cumulants are similar to moments. The difference among cumulants and moments is that the moments of a random process are derived from the characteristic function of the random variable, while the cumulant generating function is defined as the logarithm of the characteristic function of the random variable. The main properties of the cumulants are presented in appendix B of this book.

The cumulant of order n of a stationary random process, $X(k)$, can be written as [34]

$$c_n^X(\tau_1, \tau_2, \dots, \tau_{n-1}) = m_n^X(\tau_1, \tau_2, \dots, \tau_{n-1}) - m_n^G(\tau_1, \tau_2, \dots, \tau_{n-1}), \tag{2.8}$$

where $m_n^G(\tau_1, \tau_2, \dots, \tau_{n-1})$ is the nth moment of a stochastic process generated by a Gaussian distribution with a mean which equals the mean of the unknown process which generated $X(k)$ and the same autocorrelation function. That is, it is the moment of order of n of an equivalent stochastic process associated to a Gaussian distribution.

From equation (2.8) it is evident that, for a $X(k)$ generated by a stochastic process with Gaussian distribution, the cumulant for orders greater than 2 is equal to zero, since $m_n^X(\tau_1, \tau_2, \dots, \tau_{n-1}) = m_n^G(\tau_1, \tau_2, \dots, \tau_{n-1})$, and then $c_n^x(\tau_1, \tau_2, \dots, \tau_{n-1}) = 0$, for $n = 3, 4, \dots$ [34].

Notice that the cumulants can be written as arrays with entries corresponding to τ_1, τ_2, etc. For example, a cumulant of order 3 is

$$c_3^X(\tau_1, \tau_2) = E\left\{X(k)X(k + \tau_1)^*X(k + \tau_2)\right\},$$

which is a summatory (it would be an integral for continuous signals, which we do not consider here). This is a matrix. For $n > 3$ it turns into higher-dimensional arrays. For all cumulants of order n we call the origin when $\tau_i = 0$ for all $i = 1, \dots, n - 1$. For higher-order cumulants a null value at the origin does not imply that the cumulants for any point in the multidimensional space are null. Thus, for example, a null value of the asymmetry coefficient does not imply that the third-order cumulants are identically equal to zero [36]. Although fourth-order cumulants imply a considerable increase in computational complexity and cost, they are especially necessary when third-order cumulants cancel out for symmetrically distributed processes, such as uniformly distributed processes in an interval $[-a, a]$ with $a \in \mathbb{R}$, Laplace processes, and Gaussian processes. We note here that Laplace processes are those whose associated stochastic process is generated from a Laplace distribution (also called a double exponential distribution). Third-order cumulants do not cancel for processes whose probabilistic density function is not symmetric, such as exponential or Rayleigh processes (stochastic processes generated from an exponential or Rayleigh distribution, respectively), but they can take extremely small values compared with the values presented by their fourth-order cumulants [36].

2.3.2 Higher-order spectra

The spectrum of nth order of a random vector $\{X(k)\}$ is defined as the multidimensional Fourier transform $F_k[\,\cdot\,]$ of order $n - 1$. For a generic order n, the nth-order moment spectrum is defined as [37]

$$M_n^X(\omega_1, \omega_2, \dots, \omega_{n-1}) = F_n[m_n^X(\tau_1, \tau_2, \dots, \tau_{n-1})], \tag{2.9}$$

and similarly, the nth-order cumulant spectrum is defined as [38]

$$C_n^X(\omega_1, \omega_2, \dots, \omega_{n-1}) = F_n[c_n^X(\tau_1, \tau_2, \dots, \tau_{n-1})]. \tag{2.10}$$

See appendix B for a discussion on higher-order moments and cumulants. Note that the nth-order cumulant spectrum is periodic with period 2π if $X(t)$ is periodic (as assumed here). That is,

$$C_n^X(\omega_1, \omega_2, \dots, \omega_{n-1}) = C_n^X(\omega_1 + 2\pi, \omega_2 + 2\pi, \dots, \omega_{n-1} + 2\pi). \tag{2.11}$$

When working with stochastic processes, such as speech signals or ambient noise, using the cumulant spectrum has a number of advantages over the moment spectrum [34]. These are the following:

1. For Gaussian processes all cumulants of order greater than 2 cancel out. Then, a non-zero higher-order cumulant spectrum indicates non-Gaussian characteristics of a particular process.
2. For the case of a non-Gaussian process, its covariance function corresponds to the impulse function (Dirac delta in time), and, therefore, it has a flat spectrum. Its higher-order cumulants have the form of a multidimensional impulse function, and the polyspectra of this noise are multidimensionally flat.
3. The cumulant of the sum of two statistically independent random processes corresponds to the sum of the cumulants of each individual process, unlike the moments that do not satisfy this property.

Let us explicitly write the Fourier transform of cumulant sequence of order n for a discrete $X(k)$ random vector, which is [36]

$$C_n^X(\omega_1, \omega_2, \ldots, \omega_{n-1}) = \frac{1}{(2\pi)^{n-1}} \sum_{\tau_1=-\infty}^{\infty} \cdots \sum_{\tau_{n-1}=-\infty}^{\infty} c_n^X(\tau_1, \tau_2, \ldots, \tau_{n-1}) e^{-j(\omega_1\tau_1+\cdots+\omega_{n-1}\tau_{n-1})}. \quad (2.12)$$

In the following, we focus on the cumulant spectra for the cases of the power spectrum (with $n = 2$), the bispectrum (with $n = 3$), and the trispectrum (with $n = 4$).

Power spectrum
The power spectrum is obtained for $n = 2$, that is,

$$C_2^X(\omega) = \frac{1}{2\pi} \sum_{\tau=-\infty}^{\infty} c_2^X(\tau) e^{-j\omega\tau}, \quad (2.13)$$

where $|\omega| \leqslant \pi$. Notice that the cumulant of order 2 in the time domain, $c_2^X(\tau)$, is the covariance sequence of the process when the process is a zero mean stochastic process (e.g. the stochastic process is generated by additive Gaussian noise with zero mean). The expression (2.13) is also known as the Wiener–Khintchine theorem [37, 38].

Bispectrum
The bispectrum is obtained for $n = 3$, that is,

$$C_3^X(\omega_1, \omega_2) = \frac{1}{(2\pi)^2} \sum_{\tau_1=-\infty}^{\infty} \sum_{\tau_2=-\infty}^{\infty} c_3^X(\tau_1, \tau_2) e^{-j(\omega_1\tau_1+\omega_2\tau_2)}$$
$$|\,\omega_1\,| \leqslant \pi, \; |\,\omega_2\,| \leqslant \pi, \; |\,\omega_1 + \omega_2\,| \leqslant \pi \quad (2.14)$$

where $c_3^X(\tau_1, \tau_2)$ represents the sequence of third-order cumulants of $X(k)$. The cumulant expression satisfies the following symmetry relations [37]:

$$\begin{aligned} c_3^X(\tau_1, \tau_2) &= c_3^X(\tau_2, \tau_1) = c_3^X(-\tau_2, \tau_1 - \tau_2) = c_3^X(\tau_2 - \tau_1, -\tau_1) \\ &= c_3^X(\tau_1 - \tau_2, -\tau_2) = c_3^X(-\tau_1, \tau_2 - \tau_1). \end{aligned} \quad (2.15)$$

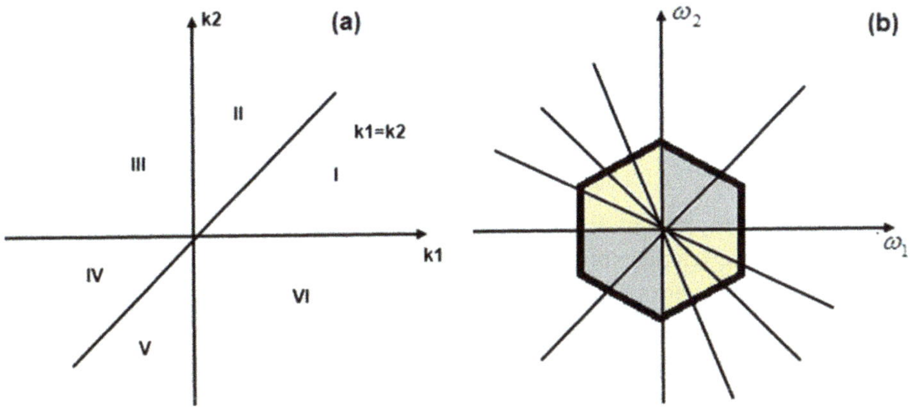

Figure 2.10. Symmetry regions for (a) third-order cumulants; (b) bispectrum.

These relations translate in a division of the plane $\tau_1\tau_2$ in six regions. Knowing the third-order cumulant in any of these six regions (see figure 2.10), the complete sequence corresponding to the third-order cumulants can be reconstructed. Note that each of these regions contains its boundary. Thus, for example, sector I is an infinite region characterized by $0 < \tau_2 \leqslant \tau_1$ ($\pi/4$ sector belonging to the first quadrant). We emphasize that for non-stationary processes, these six regions of symmetry disappear. This one of the reasons why we restrict ourselves in this book to stationary processes. Nevertheless it can be applied to non-stationary processes under certain conditions [39, 40].

From these relations and the definition of the third-order cumulant spectrum, the following relations in the two-dimensional frequency domain are obtained [35]:

$$\begin{aligned} C_3^X(\omega_1, \omega_2) &= C_3^X(\omega_2, \omega_1) = C_3^X(-\omega_2, -\omega_1) = C_3^X(-\omega_1 - \omega_2, \omega_2) \\ &= C_3^X(\omega_1, -\omega_1 - \omega_2) = C_3^X(-\omega_1 - \omega_2, \omega_1) = C_3^X(\omega_2, \omega_1 - \omega_2) \end{aligned} \tag{2.16}$$

Figure 2.10(b) represents the 12 bispectrum symmetry regions when real stochastic processes are considered and, analogously to the time domain, the knowledge of the bispectrum in the triangular region $\omega_2 \geqslant 0$, $\omega_1 \geqslant \omega_2$, $\omega_1 + \omega_2 \leqslant \pi$ is sufficient for a complete reconstruction of the bispectrum. Note that, in the frequency domain, the symmetry regions present a finite area and in them, in general, the bispectrum takes complex values and, consequently, the phase information is preserved [35]. This is an advantage over classical methods (see [35]) like second-order statistics (autocovariance function) where the phase information is lost. Likewise in this book a new phase recovery method is proposed to reduce the computational cost that implies the bispectrum calculation as multidimensional function (see appendix B).

Trispectrum

The trispectrum is obtained for $n = 4$, that is,

$$C_4^X(\omega_1, \omega_2, \omega_3) = \frac{1}{(2\pi)^3} \sum_{\tau_1=-\infty}^{\infty} \sum_{\tau_2=-\infty}^{\infty} \sum_{\tau_3=-\infty}^{\infty} c_4^X(\tau_1, \tau_2, \tau_3) e^{-j(\omega_1\tau_1+\omega_2\tau_2+\omega_3\tau_3)}, \qquad (2.17)$$

with

$$|\omega_1| \leqslant \pi, |\omega_2| \leqslant \pi, |\omega_3| \leqslant \pi |, \omega_1 + \omega_2 + \omega_3| \leqslant \pi, \qquad (2.18)$$

where $c_4^X(\tau_1, \tau_2, \tau_3)$ represents the sequence of fourth-order cumulants. By combining the definition of the trispectrum and that of the fourth-order cumulants, 96 regions of symmetry can be deduced when evaluating real processes. From the spectra of higher-order cumulants, the expressions of their respective cumulants in the time domain can be recovered by applying the inverse Fourier transform of order n (see [36, 37]).

2.3.3 Use of HOSA applied to noise reduction

The use of HOSA methods for noise reduction has proven to be very effective [37]. This is due to two fundamental properties that cumulants possess:

- The higher-order cumulant is zero for a random process with normal distribution.
- The cumulant of a sum of independent signals is equivalent to the sum of the cumulants of each individual signal.

Several works have been carried out in this direction, focusing in general on noise reduction for single-input systems, applications of signal detection, system identification, and restoration of images degraded by random signals of Gaussian nature [41].

Several investigations have been carried out and various algorithms have been developed focusing, in particular, on the noise reduction problem. These works focus on the use of HOSA to perform the task of noise reduction in signals. In [41] a triple cross-correlation is used as a way to minimize the error in a noise reduction system following the mean square error (MSE) criterion when the signal is contaminated by noise of Gaussian nature with unknown covariance function. This work is designed for its use in adaptive systems since it requires an additional reference signal. Its novelty lies in the use of statistical analysis, in this case cross-correlation, to minimize the cost function described by Widrow *et al* in [42]. In [43] the development of Wiener filters based on the use of cumulants for error minimization using the MSE criterion when the noise-free input signal and the reference signal are non-Gaussian and contaminated by normally distributed noise is discussed; moreover, it is based on the fact that the cumulants of processes of Gaussian nature, of orders greater than 2, are null. In [44, 45] an algorithm is developed for the identification of linear systems in which both the input and output are contaminated by noise with Gaussian distribution. These works use higher-order statistics for this identification based on the third-order cumulant; the whole algorithm is integrated to an adaptive identification system that uses the RLS algorithm as an update criterion. In [46] the use of first-, second-, and third-order moments to reduce noise from rotating machines in a scheme using reference signals is reported. Also in [47] the third-order moments of the data observation are used to evaluate the process of filter adaptation, instead of the classical Wiener filter approach, in a noise reduction system. The use of adaptive filters using higher-order statistics for noise reduction instead of the use of the MSE criterion based on second-order statistics is shown in [48, 49].

All previous works based on adaptive systems are limited by the use of the additional reference signal to perform the noise attenuation task. Moreover, they are only applied to random processes of Gaussian nature. Although they present as a novelty the use of HOSA as an error minimization criterion in adaptive systems, an improvement based on the signal asymmetry coefficient is proposed in [49].

In turn, the use of HOSA has been proposed in the analysis of polyspectra as a form of noise attenuation in signals. In [50] the development of a method to minimize the third-order cumulant of a signal is proposed, taking into account that theoretically this becomes zero. The method is based on minimizing a function that represents the power of the signal in the bispectrum of the analyzed data sequence and whose objective is to improve the results of the third-order statistics in a series of random data with symmetric probabilistic density function.

In [51] the approach of higher-order statistics and analysis through polyspectra to obtain consistent nonparametric estimators of the autocorrelation and spectrum in a signal, instead of the use of second-order statistical estimators, is discussed. In [52], an algorithm based on the bispectrum to obtain the waveform of a signal in a series of data contaminated with variable delay, for example in a multiple-input–multiple-output system, is shown. A comparison with several algorithms in terms of obtaining the amplitude and phase information is made therein.

Also in [53, 54] respectively, the application of the bispectrum in the reconstruction of images degraded by stationary noise with normal distribution and multiplicative noise is discussed.

In [55] the authors approach the study of the estimation of signals contaminated by noise with non-Gaussian distribution and unknown covariance function, i.e. the opposite problem to the classical use of HOSA for the estimation and detection of signals in the presence of Gaussian noise, and propose a solution based on the calculation of cumulants.

An alternative solution for the treatment of noise is used in [56], where a method for the separation of the sum of two non-Gaussian data series is proposed; the work is based on the use of the power spectrum and the fourth-order spectrum. In [57] a generic model of probabilistic noise density distribution is proposed to optimize detection in non-stationary environments, using higher-order statistics through the use of kurtosis. All previous works related to the use of higher-order spectra in noise attenuation tasks are limited by the computational complexity involved in the calculation of two-dimensional functions.

Likewise, they only seek to perform a polyspectral analysis of the useful signal that is immune to the influence of the contaminating noise; they do not obtain the waveform of the signal at the output, and their application is focused on noise of a Gaussian nature.

2.3.4 Use of HOSA applied to phase information retrieval

The use of HOSA for phase information retrieval has been discussed in several papers. Some methods developed for phase retrieval only are based on higher-order spectra [58–63], using the criterion that polyspectra preserve phase information. However, the tools proposed in these works exhibit a high computational cost since the higher-order parameters impose the calculation of multidimensional functions, which makes these methods a non-viable solution for practical use, besides not being directly applied to the general problem of obtaining the amplitude, frequency, and phase of a signal contaminated by noise. They are only framed in obtaining the phase information from the vector obtained at the output of the higher-order spectrum calculation.

On the other hand, [64] refers to a method based on Bayes' theorem and the Fourier transform for phase recovery, in addition to using the function interpolation criterion for information reconstruction, but its computational complexity is high.

In order to obtain an efficient solution, methods based on partial regions of the higher-order spectra have been developed [65–67], but they only approach the phase recovery problem in linear and time invariant systems and do not take into account the other signal parameters, and their computational cost is high since they use the two-dimensional Fourier transform.

All previous works related to the use of higher-order spectra or partial regions of these in tasks of obtaining phase information are limited by the computational complexity involved in the calculation of two-dimensional functions.

Moreover, they are only based on obtaining the phase in a similar way to the works previously referred to regarding noise attenuation; the output signal waveform is not obtained. We detail a solution to this issue in the next chapter.

2.3.5 Conclusions of chapter

In this chapter we have reviewed several of the digital signal processing methods applied to noise reduction, in addition to showing some examples in certain application areas, which evidence the wide range of current methodologies and conceptions of signal processing, depending on the environment, the process, and the nature of the signal for each particular application. In addition, a classification has been proposed for the different noise reduction techniques currently used, and they were explained according to the system design and processing conditions imposed by each of the methods discussed.

Furthermore, the fundamentals of HOSA have been shown, for which an introduction to these methods, very useful when determining specific characteristics in processes affected by disturbing signals, has been given.

On top of this, a description of the referenced works using HOSA for noise reduction and phase information retrieval has been given.

One of the potential applications of higher-order statistical processing is the task of noise attenuation, due to the advantages offered by this type of processing using the statistical tools known as cumulants, taking into account that the cumulant for orders greater than 2, of a Gaussian distorting sample, is zero.

Although this represents a higher computational cost than the alternative of using classical processing techniques, depending on the application, this cost may be justified by the high effectiveness obtained.

References

[1] Vaseghi S V 2000 *Advanced Digital Signal Processing and Noise Reduction* 2nd edn (New York: Wiley)

[2] Farhang-Boroujeny B 1998 *Adaptive Filters: Theory and Applications* (New York: Wiley)

[3] Trimale M B and Chilveri 2017 A review: FIR filter implementation *2nd IEEE Int. Conf. on Recent Trends in Electronics, Information Communication Technology (RTEICT)* pp 137–41

[4] Dwivedi A K, Ghosh S and Londhe N D 2018 Review and analysis of evolutionary optimization-based techniques for FIR filter design *Circuits Syst. Signal Process.* **37** 4409–30

[5] Lu L, Yin K-L, de Lamare R C, Zheng Z, Yu Y, Yang X and Chen B 2021 A survey on active noise control in the past decade–part I: linear systems *Signal Process.* **183** 108039

[6] Lu L, Yin K-L, de Lamare R C, Zheng Z, Yu Y, Yang X and Chen B 2021 A survey on active noise control in the past decade–part II: nonlinear systems *Signal Process.* **181** 107929

[7] Kumar K, Pandey R, Karthik M L N S, Bhattacharjee S S and George N V 2021 Robust and sparsity-aware adaptive filters: a review *Signal Process.* **189** 108276

[8] Arsri S W, Wahyunggoro O and Cahyadi A I 2022 A review of adaptive filter algorithm-based battery state of charge estimation *2021 Int. Seminar on Machine Learning, Optimization, and Data Science (ISMODE)* pp 125–30

[9] Rusu A-G, Ciochină S, Paleologu C and Benesty J 2022 Cascaded RLS adaptive filters based on a Kronecker product decomposition *Electronics* **11** 409

[10] Bejani M M and Ghatee M 2021 A systematic review on overfitting control in shallow and deep neural networks *Artif. Intell. Rev.* **54** 6391–438

[11] Song H, Kim M, Park D, Shin Y and Lee J-G 2022 Learning from noisy labels with deep neural networks: a survey *IEEE Trans. Neural Netw. Learn. Syst.* accepted

[12] Salimy A, Mitiche I, Boreham P, Nesbitt A and Morison G 2022 Dynamic noise reduction with deep residual shrinkage networks for online fault classification *Sensors* **22** 515

[13] Zubaidi S L, Al-Bugharbee H, Ortega-Martorell S, Gharghan S K, Olier I, Hashim K S, Al-Bdairi N S S and Kot P 2020 A novel methodology for prediction urban water demand by wavelet denoising and adaptive neuro-fuzzy inference system approach *Water* **12** 1628

[14] Diao L, Niu D, Zang Z and Chen C 2019 Short-term weather forecast based on wavelet denoising and catboost *2019 Chinese Control Conf. (CCC)* pp 3760–4

[15] Zeng Z and Khushi M 2020 Wavelet denoising and attention-based RNN-ARIMA model to predict forex price *2020 Int. Joint Conf. on Neural Networks (IJCNN)*

[16] Guo J, Zhang H, Zhen D, Shi Z, Gu F and Ball A D 2020 An enhanced modulation signal bispectrum analysis for bearing fault detection based on non-Gaussian noise suppression *Measurement* **151** 107240

[17] Yang L, Zhen P, Wang J, Zhang J and Guo D 2021 Radar emitter recognition method based on bispectrum and improved multi-channel convolutional neural network *6th Int. Conf. on Communication, Image and Signal Processing (CCISP)* pp 321–8

[18] Liang S, Hu T, Qiao B and Cui D 2022 Reflection seismic interferometry via higher-order cumulants to solve normal moveout stretch *IEEE Geosci. Remote. Sens. Lett.* **19** 1–5
[19] Mitra S K and Kaiser J F 1993 *Handbook for Digital Signal Processing* (New York: Wiley)
[20] Sayed Ali H 2003 *Fundamentals of Adaptive Filtering* (New York: Wiley)
[21] Martin del Brío Bonifacio and Sanz Alfredo 2001 *Redes Neuronales y Sistemas Borrosos* 2nd edn (Mexico City: Alfaomega Ra-Ma)
[22] Bagheri F, Ghafarnia N and Bahrami F 2013 Electrocardiogram (ECG) signal modeling and noise reduction using Hopfield neural networks *Eng. Technol. Appl. Sci. Res.* **3** 345–8
[23] Badri L 2010 Development of neural networks for noise reduction *Int. Arab. J. Inf. Technol.* **7** 289–94
[24] Chen D 2022 Artificial-Neural-Network-MATLAB. GitHub https://github.com/drBearcub/Artificial-Neural-Network-MATLAB
[25] Chaudhary D 2022 Neural-Networks-MATLAB. GibHub https://github.com/darshanime/neural-networks-MATLAB
[26] Moghadas D 2022 ANN-Based-Forward. GitHub https://github.com/ML4Geophysics/ANN-Based-Forward
[27] Baleanu D (ed) 2012 *Advances in Wavelet Theory and Their Applications in Engineering* (London: IntechOpen)
[28] Cohen R 2012 *Signal Denoising Using Wavelets* Department of Electrical Engineering Technion, Israel Institute of Technology, Haifa
[29] Comon P and Pham D T 1990 Estimating the order of a FIR filter for noise cancellation *IEEE Trans. Inf. Theory* **36** 429–34
[30] Ouahabi A 2013 A review of wavelet denoising in medical imaging *8th Int. Workshop on Systems, Signal Processing and their Applications (WoSSPA)* pp 19–26
[31] Carter B and Mancini R 2009 *Op Amps for Everyone* (Dallas, TX: Texas Instruments)
[32] Wang Q-B, Yang Y-J and Zhang X 2020 Weak signal detection based on Mathieu-Duffing oscillator with time-delay feedback and multiplicative noise *Chaos Solitons Fractals* **137** 109832
[33] Shah A, Bangash J I, Khan A W, Ahmed I, Khan A, Khan A and Khan A 2022 Comparative analysis of median filter and its variants for removal of impulse noise from gray scale images *J. King Saud Univ. - Comput. Inf. Sci.* **34** 505–19
[34] Boashash B, Powers E J and Zoubir A M 1995 *Higher-Order Statistical Signal Processing* (New York: Wiley) pp 27–48
[35] Mendel J M 1991 Tutorial on higher-order statistics (spectra) in signal processing and system theory: theoretical results and some applications *Proc. IEEE* **79** 278–305
[36] Kay S M 1993 *Fundamentals of Statistical Signal Processing, Estimation Theory* (Englewood Cliffs, NJ: Prentice-Hall)
[37] Swami A, Giannakis G B and Zhou G 1997 Bibliography on higher-order statistics *Signal Process.* **60** 65–126
[38] Howard R M 2009 *Principles of Random Signal Analysis and Low Noise Design* (New York: Wiley)
[39] Vosoughi E and Javaherian A 2018 Parameters effective on estimating a nonstationary mixed-phase wavelet using cumulant matching approach *J. Appl. Geophys.* **148** 83–97
[40] Chen B, Guo R, Yu S and Yu Y 2021 An active noise control method of non-stationary noise under time-variant secondary path *Mech. Syst. Signal Process.* **149** 107193

[41] Dandawate A V and Giannakis G B 1989 A triple cross-correlation approach for enhancing noisy signals *Workshop on Higher-Order Spectral Analysis* 212–6
[42] Widrow B, Glover J R and McCool J M 1975 Adaptive noise cancelling: principles and applications *Proc. IEEE* **63** 1692–716
[43] Feng C-C and Chi C-Y 1996 Design of Wiener filters using a cumulant based MSE criterion *Signal Process.* **54** 23–48
[44] Giannakis G B and Dandawate A V 1990 Linear and non-linear adaptive noise cancelers *Int. Conf. on Acoustics, Speech, and Signal Processing* **3** 1373–6
[45] Giannakis G B and Dandawate A V 1991 Higher-order statistics based input/output system identification and application to noise cancellation *Circuits Syst. Signal Process.* **10** 485–511
[46] Serviere C and Baudois D 1993 Noise reduction with sinusoidal signals *IEEE Signal Processing Workshop on Higher-Order Statistics* 230–4
[47] Serviere C, Baudois D and Silvent A 1991 Estimation of correlated frequencies in noise canceling using higher order moments *Int. Conf. on Acoustics, Speech, and Signal Processing* **5** 3477–80
[48] Shin D C and Nikias C L 1994 Adaptive interference canceler for narrowband and wideband interferences using higher order statistics *IEEE Trans. Signal Process.* **42** 2715–28
[49] Iglesias Martínez M E 2013 Analysis and implementation of LMK and LM-skewness algorithms on FPGA *Application to Noise Cancellation, IV Argentine Conf. on Embedded Systems (14–16 August 2013)* (CONICET, ACSE)
[50] Durrani T S, Leyman A R and Soraghan J J 1994 New cumulant suppression technique *Electron. Lett.* **30** 623–4
[51] Giannakis G B and Delopoulos A N 1990 Nonparametric estimation of autocorrection and spectra using cumulants and polyspectra *Proc. SPIE* **1348** 503–17
[52] Nakamura M 1993 Waveform estimation from noisy signals with variable signal delay using bispectrum averaging *IEEE. Trans. Biomed. Eng.* **40** 118–27
[53] Newman J D and Van Vranken R C 1990 Shift-invariant imaging through high background noise using the bispectrum *Proc. ASSP Workshop on Spectrum Estimation and Modeling (Rochester, NY)*
[54] Raghuveer M R, Wear S and Song J 1991 Restoration of speckle degraded images using bispectra *Int. Conf. on Acoustics, Speech, and Signal Processing (ICASSP 91)* pp 3077–80
[55] Sadler B M, Giannakis G B and Lii K S 1994 Estimation and detection in the presence of non-Gaussian noise *IEEE Trans. Signal Process.* **42** 2729–41
[56] Sakaguchi F and Sakai H 1987 A spectrum separation method for the sum of two non-Gaussian time series using higher order periodograms *IEEE J. Ocean. Eng.* **12** 80–9
[57] Tesei A and Regazzoni C S 1996 The asymmetric generalized Gaussian function: a new HOS-based model for generic noise PDFs *Proc. 8th Workshop on Statistical Signal and Array Processing* pp 210–3
[58] Bartelt H, Lohmann A W and Wirnitzer B 1984 Phase and amplitude recovery from bispectra *Appl. Opt.* **23** 3121–9
[59] Matsuoka T and Ulrych T J 1984 Phase estimation using the bispectrum *Proc. IEEE* **72** 1403–11
[60] Petropulu A P and Nikias C L 1992 Signal reconstruction from the phase of bispectrum *IEEE Trans. Signal Process.* **40** 601–10
[61] Holambe R S, Ray A K and Basu T K 1996 Signal phase recovery using the bispectrum *Signal Process.* **55** 321–37

[62] Pan R and Nikias C L 1987 Phase reconstruction in the trispectrum domain *IEEE Trans. Signal Process.* **35** 895–7
[63] Dianat S A and Raghuveer M R 1990 Fast algorithms for bispectral reconstruction of two-dimensional signals *Int. Conf. Acoustics, Speech, and Signal Processing* pp 2377–9
[64] Sacchi M D, Ulrych T J and Walker C J 1998 Interpolation and extrapolation using a high-resolution discrete Fourier transform *IEEE Trans. Signal Process.* **46** 31–8
[65] Kachenoura A, Albera L, Bellanger J-J and Senhadji L 2008 Nonminimum phase identification based on higher order spectrum slices *IEEE Trans. Signal Process.* **56** 1821–9
[66] Petropulu A P and Pozidis H 1998 Phase reconstruction from bispectrum slices *IEEE Trans. Signal Process.* **46** 527–30
[67] Pozidis H and Petropulu A P 1998 System reconstruction based on selected regions of discretized higher order spectra *IEEE Trans. Signal Process.* **46** 3360–77

IOP Publishing

Algorithms for Noise Reduction in Signals
Theory and practical examples based on statistical and convolutional analysis
Miguel Enrique Iglesias Martínez, Miguel Ángel García March,
Carles Milián Enrique and Pedro Fernández de Córdoba

Chapter 3

Noise reduction in periodic signals based on statistical analysis

The higher-order cumulant of a signal contaminated by normally distributed noise is equal to the higher-order cumulant of the signal without noise. On this subject several works have focused in the approach to the problem of harmonic recovery (see chapter 2 and also [1–3]). However, these works only focus on the recovery of the amplitude of each harmonic; the phase information is not recovered. Other investigations, such as [4] and [5], use higher-order (fourth-order cumulant) statistics for the adaptive estimation of the amplitude parameters of the harmonic components of real zero mean signals.

In order to obtain a general algorithm for the estimation of periodic signals contaminated by noise of a stationary nature that groups in a single procedure the obtaining of all the parameters (amplitude and phase) of any periodic signal, a new method based on joint techniques of higher-order statistical analysis and convolution is proposed in [6].

We introduce and detail in this chapter the theoretical foundations of the use of higher-order statistics for noise reduction tasks, as well as a comparative analysis with respect to the other existing phase recovery methods based on higher-order statistics. Also, the basic principles of statistical signal processing applied to noise reduction will be discussed, as well as the theoretical description of the proposed algorithm. In addition we also show the experimental results obtained from the simulation point of view and working with real, previously digitized signals.

3.1 Basic approach to noise reduction using higher-order noise reduction statistics

For a given harmonic real signal, the observed data can be described as follows [2, 7]:

$$y(t) = \sum_{k=1}^{N} A_k \cos\left(2\pi f_k t + \phi_k\right) + n(t) = x(t) + n(t), \tag{3.1}$$

doi:10.1088/978-0-7503-3591-1ch3

where $x(t)$ is the useful signal (signal to be detected) and $n(t)$ corresponds to the additive noise of zero mean value, with probabilistic Gaussian distribution function. Here, A_k, f_k, and ϕ_k are the amplitude, frequency, and phase values, respectively, of the harmonic signal, which are not known.

As previously mentioned, the second-order cumulant of a Gaussian noise process is non-zero. For this reason, to reduce noise it is more convenient to estimate the higher-order cumulant, for example the third-order cumulant for the process described in (2.1), which can be defined as follows:

$$c_3^y(\tau_1, \tau_2) = c_3^x(\tau_1, \tau_2) + c_3^n(\tau_1, \tau_2). \tag{3.2}$$

However, the calculation of the third-order cumulant for signals with symmetric probabilistic density function is zero [8–10]. Consequently, the third-order cumulant of a harmonic signal is zero. For this reason in this research we will work with the estimation of the fourth-order cumulant.

3.1.1 Working with the fourth-order cumulant

As discussed in chapter 2, for a stationary random process $z(t)$ of zero mean value and for $k = 3, 4, \ldots$, the cumulant of order k of $z(t)$ can be defined in terms of moments as [8, 10]

$$c_k^z(\tau_1, \tau_2, \ldots, \tau_{k-1}) = E\{z(\tau_1)\cdots z(\tau_{k-1})\} - E\{g(\tau_1)\cdots g(\tau_{k-1})\}, \tag{3.3}$$

where $g(t)$ is a Gaussian process with the same second-order statistics as $z(t)$. Then, if $z(t)$ is Gaussian, the higher-order cumulants are all zeros. The fourth-order cumulant of the process described in equation (3.1) can be calculated as follows:

$$c_4^y(\tau_1, \tau_2, \tau_3) = c_4^x(\tau_1, \tau_2, \tau_3) + c_4^n(\tau_1, \tau_2, \tau_3). \tag{3.4}$$

According to the process described in equation (3.1), if $n(t)$ is a Gaussian random signal of zero mean value, $c_4^n(\tau_1, \tau_2, \tau_3) = 0$, then $c_4^y(\tau_1, \tau_2, \tau_3) = c_4^x(\tau_1, \tau_2, \tau_3)$. For zero mean Gaussian processes the fourth-order cumulant can be calculated as follows [10, 11]:

$$\begin{aligned} c_4^x(\tau_1, \tau_2, \tau_3) = {} & E\{x(t) \cdot x(t + \tau_1) \cdot x(t + \tau_2) \cdot x(t + \tau_3)\} \\ & - c_2^x(\tau_1) \cdot c_2^x(\tau_2 - \tau_3) - c_2^x(\tau_2) \cdot c_2^x(\tau_3 - \tau_1) \\ & - c_2^x(\tau_3) \cdot c_2^x(\tau_1 - \tau_2). \end{aligned} \tag{3.5}$$

If one works only with the one-dimensional component of the fourth-order cumulant, taking $\tau_2 = \tau_3 = 0$, one arrives at the same result as in [12, 13] (and which was obtained in a very similar way in [14], assuming $\tau_1 = \tau_2 = \tau_3 = \tau$). This one-dimensional component obtained from equation (3.5), assuming $\tau_1 = \tau_2 = \tau_3 = \tau$, lies on equation (3.6), and with this, two of the useful signal parameters can be recovered: amplitude and frequency. Although the phase is not preserved, the noise is completely removed. This component, of smaller calculation, can be conveniently used when implementing this method in hardware platforms, such as field programmable gate arrays (FPGAs), microcontrollers, and digital signal processing, among others:

$$c_4^x(\tau_1, 0, 0) = E\{x(t)^3 \cdot x(t + \tau_1)\} - 3 \cdot E\{x(t) \cdot x(t + \tau_1)\} \cdot E\left\{x^2(t)\right\}. \quad (3.6)$$

Developing the left term of the subtraction in equation (3.6) and applying the trigonometric identity $\cos A \cdot \cos B = \frac{1}{2}\cos(A - B) + \frac{1}{2}\cos(A + B)$, we obtain the following (see appendix C):

$$\begin{aligned} E\left\{x(t)^3 \cdot x(t + \tau_1)\right\} &= E\left\{\sum_{k=1}^{N} A_k \cos(w_k t + \phi_k)^3 \cdot \sum_{k=1}^{N} A_k \cos(w_k t + w_k\tau_1 + \phi_k)\right\} \\ &= \sum_{k=1}^{N} A_k^4 E\left\{\cos(w_k t + \phi_k)^3 \cdot \cos(w_k t + w_k\tau_1 + \phi_k)\right\} \\ &= \sum_{k=1}^{N} \frac{3A_k^4}{8}\cos(w_k\tau_1). \end{aligned} \quad (3.7)$$

The last term in equation (3.6) corresponds to the autocorrelation function of a harmonic signal multiplied by the variance of the signal itself. We summarize here (see also appendix D) the procedure used to deal with this term:

$$\begin{aligned} 3 \cdot E\{x(t) \cdot x(t + \tau_1)\} \cdot E\{x^2(t)\} &= 3 \cdot E\left\{\sum_{k=1}^{N} A_k \cos(w_k t + \phi_k) \cdot \sum_{k=1}^{N} A_k \cos(w_k t + w_k\tau_1 + \phi_k)\right\} \\ &\quad \cdot E\{x^2(t)\} \\ &= 3 \cdot \sum_{k=1}^{N} A_k^2 E\{\cos(w_k t + \phi_k) \cdot \cos(w_k t + w_k\tau_1 + \phi_k)\} \cdot E\{x^2(t)\} \\ &= 3 \cdot \sum_{k=1}^{N} \frac{A_k^2}{2}\cos(w_k\tau_1) \cdot E\{x^2(t)\} \\ &= 3 \cdot \sum_{k=1}^{N} \frac{A_k^2}{2}\cos(w_k\tau_1) \cdot \sum_{k=1}^{N} \frac{A_k^2}{2} \\ &= \sum_{k=1}^{N} \frac{3A_k^4}{4}\cos(w_k\tau_1). \end{aligned} \quad (3.8)$$

Substituting the results obtained in equations (3.7) and (3.8), we obtain

$$\begin{aligned} c_4^x(\tau_1, 0, 0) &= \sum_{k=1}^{N} \frac{3A_k{}^4}{8}\cos(w_k\tau_1) - \sum_{k=1}^{N} \frac{3A_k{}^4}{4}\cos(w_k\tau_1) \\ &= \sum_{k=1}^{N} -\frac{3A_k{}^4}{8}\cos(w_k\tau_1). \end{aligned} \quad (3.9)$$

As can be seen in equation (3.9), the original spectral components of the signal are obtained, but without noise, and with an amplitude that is a function of the original amplitude. However, the original signal $x(t)$ is not obtained; this is due to the loss of the phase information during the noise reduction process, a problem that will be discussed later.

3.1.2 Experimental results on noise reduction applying only higher-order (fourth-order) statistics

For the practical verification of the method described in the previous section, it will be applied to a multitone signal with different frequency and phase for each of the given harmonic components. For this case, three tones of frequency, 50 Hz, 100 Hz, and 400 Hz, were generated in MATLAB, as well as corresponding phases of $\frac{\pi}{4}$, $\frac{\pi}{6}$, and $\frac{\pi}{3}$ radians, at a sampling frequency of 2.5 kHz. The interfering signal was generated from the MATLAB *randn* function. This function generates a vector of random data whose probabilistic distribution function is Gaussian. A signal window equivalent to 102 400 samples was used to test the hypothesis of the algorithm. From the total data window 100 realizations of length 1024 samples of signal plus noise were used and processed using the cumulant-based algorithm. Table 3.1 summarizes the results obtained; each column represents the average value of the 100 realizations of the processed data window.

Figure 3.1 shows the results obtained after evaluating the one-dimensional component of the fourth-order cumulant of the contaminated signal. Figure 3.1(a) shows a comparison between the output obtained after the calculation of the fourth-order cumulant $c_4^y(\tau_1, 0, 0)$ and the useful signal without noise. On the other hand, figure 2.10(b) shows a comparison of the spectrum of the signal at the output of the calculation of the one-dimensional component of the fourth-order cumulant and the spectrum of the contaminated signal. As can be seen in figure 3.1(a), the output signal waveform is different from the input signal waveform. This is due to the loss of the phase information of the components of each harmonic during the calculation of the fourth-order cumulant, which will be discussed later. However, the noise is attenuated compared with the input.

The results obtained show the loss of the original information of the signal phase. This is in accordance with the theoretical result discussed in the section above (equation (3.9)). As shown in table 3.1, the noise is reduced and the SNR at the output is increased, but there is no proper correlation of the signal obtained with respect to the input signal, due to the loss of phase information after evaluating the higher-order statistic.

3.1.3 Phase recovery algorithm

We introduce here an algorithm which permits us to perform noise reduction using the result of the fourth-order cumulant calculation without losing the phase information of the original signal. This algorithm is based on the convolution between the contaminated input signal, $y(t)$ (equation (3.1)), and the result of the

Table 3.1. Experimental noise reduction results applying only higher-order (fourth-order) statistics. Average values obtained from 100 noise realizations, for a window of 102 400 samples.

Signal	Input signal-to-noise ratio (SNR)	Output SNR	Input correlation	Output correlation
Multitone	3.988 7	9.872 1	0.846 8	−0.015 1

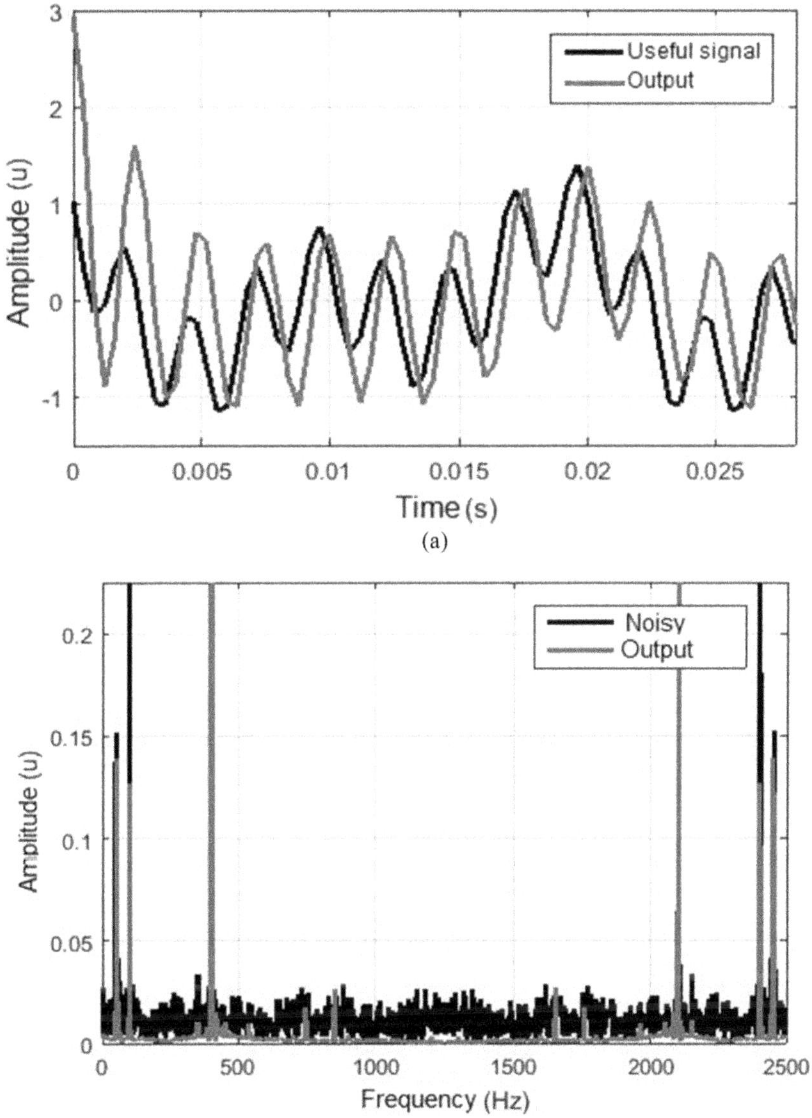

Figure 3.1. (a) Comparison between the multitone signal and the output of $c_4^y(\tau_1, 0, 0)$. (b) Comparison of the spectrum of the contaminated multitone signal and the output of $c_4^y(\tau_1, 0, 0)$.

calculation of the one-dimensional component of the fourth-order cumulant of the contaminated signal, $c_4^y(\tau_1, 0, 0)$ (equation (3.9)). Let $a(t) = y(t)$ be a harmonic signal affected by a noise process $n(t)$ with Gaussian probabilistic distribution function, and let $b(t)$ be the one-dimensional component of the fourth-order cumulant of the contaminated signal, $c_4^y(\tau_1, 0, 0) = \sum_{k=1}^{N} - \frac{3A_k^4}{8} \cos(w_k \tau_1)$. This is a periodic signal with zero phase free of noise. The proposed convolution process for phase information recovery is developed as follows. First we compute

$$\begin{aligned} a(\tau)^*b(\tau) &= \lim_{T\to\infty}\frac{1}{T}\int_{-T/2}^{T/2} a(t)\cdot b(-t+\tau)dt \\ &= \lim_{T\to\infty}\frac{1}{T}\int_{-T/2}^{T/2}\sum_{k=1}^{N}\left[A_k\cos\left(w_kt+\phi_k\right)+n(t)\right] \\ &\quad\left[-\frac{3A_k^4}{8}\cos(-w_kt+w_k\tau)\right]dt \\ &= \lim_{T\to\infty}\frac{1}{T}\int_{-T/2}^{T/2}\sum_{k=1}^{N}-\frac{3A_k^5}{8}\cos\left(w_kt+\phi_k\right)\cdot\cos\left(-w_kt+w_k\tau\right)dt \\ &\quad+\int_{-T/2}^{T/2}\sum_{k=1}^{N}-\frac{3A_k^4}{8}\cos\left(-w_kt+w_k\tau\right)\cdot n(t)dt. \end{aligned} \tag{3.10}$$

And we get

$$a(\tau)^*b(\tau)=\sum_{k=1}^{N}-\frac{3A_k^5}{16}\cos\left(w_k\tau+\phi_k\right). \tag{3.11}$$

Equation (3.11) allows one to obtain an equivalent of the original noise-free periodic signal, preserving the phase of the spectral components, with amplitudes that are a function of the original amplitudes.

3.2 Amplitude correction in the spectral domain

As can be seen in equation (3.11), the original individual amplitude of each harmonic of the deterministic periodic signal at the input of the noise reducer is affected by a factor equal to

$$A_{0_f}=\frac{3}{16}A_{i_f}^5, \tag{3.12}$$

where A_{i_f} represents the amplitude of a harmonic at frequency f of the original input periodic signal and A_{0_f} is the amplitude of this same harmonic, obtained as a result of the calculation established in equation (3.11). This represents a difficulty because the amplitude of each harmonic component in the signal $x(t)$ must be corrected independently. In this book we propose to adjust, in the spectral domain, the resulting amplitude of each harmonic component. The process involves calculating the Fourier transform of the signal resulting from the convolution process and then applying the adjustment factor from the term A_{i_f}, obtained from equation (3.12), to each spectral component individually. Once this process has been carried out, the result obtained is Fourier anti-transformed.

Two drawbacks arise from the practical experimental process of individual spectral adjustment of the amplitudes of the spectral components. The first drawback is that the noise in practice is not completely eliminated at the output of the convolution process, and, depending on the magnitude of this noise, when the spectral adjustment process is carried out, some of its spectral components may be amplified (e.g. $\sqrt[5]{0.1} = 0.63$). To reduce this difficulty in a practical way, it is proposed to calculate the mean of the resulting amplitude spectrum and subtract it from the amplitude spectrum, so that the output spectrum is adjusted with zero mean value.

The second drawback that appears when performing the individual experimental practical adjustment of the amplitudes of the spectral components has to do with the fact that the amplitude spectrum has been adjusted by a non-constant factor along f. This generates an amplitude distortion in the signal, when anti-transforming, which can be seen in figure 3.2, where the signal at the output of the proposed algorithm is shown with an amplitude distortion in the form of a non-continuous envelope.

Figure 3.2 shows the effect that the non-proportional adjustment causes f in the time domain, reflected in a gradual variation of the signal envelope. For this reason, an envelope detector is applied to the signal resulting from the anti-transformation process, and the amplitude of the 'modulated' signal is corrected by multiplying the latter by the inverse of the envelope obtained. The scheme proposed for the detection of the signal envelope is shown in figure 3.3. This scheme is based on the Hilbert transform of the input data sequence and the calculation of the absolute value of the output of the transformation process. Another scheme could also be used based on the squaring and multiplication by a gain factor of 2 of the data obtained at the

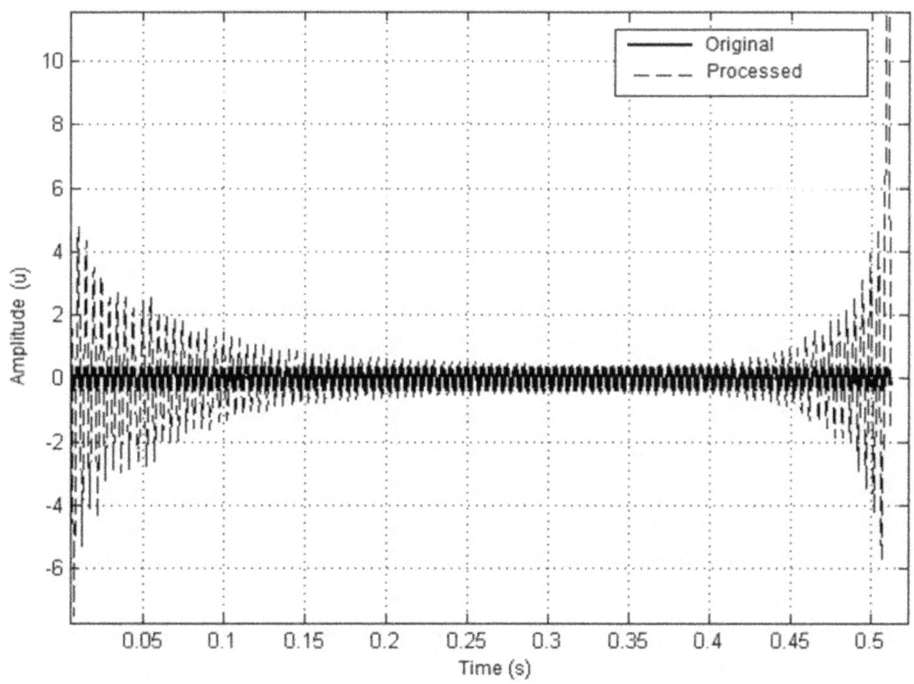

Figure 3.2. Sample signal at the output of the proposed algorithm without amplitude adjustment.

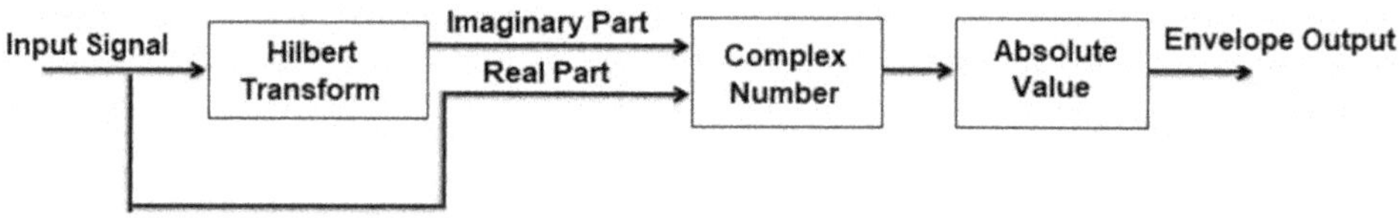

Figure 3.3. Block diagram of the envelope detector used.

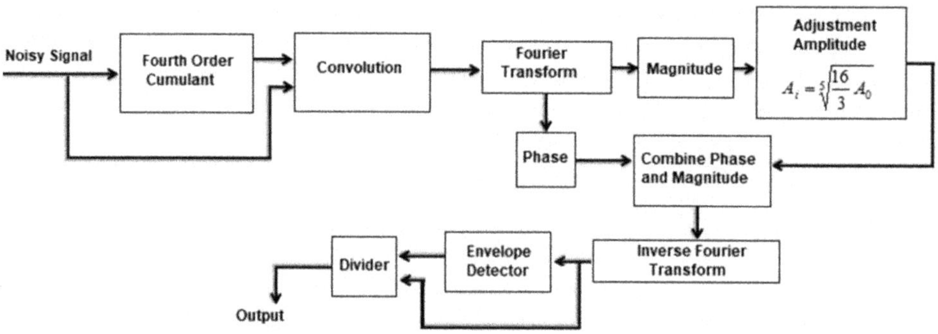

Figure 3.4. Block diagram of the proposed algorithm.

output of the inverse fast Fourier transform anti-transformation block, in addition to the use of a low pass minimum phase finite impulse response filter, based on the Parks–McClellan algorithm [11, 12, 15]. The Hilbert transform–based algorithm was chosen since the filtering-based method depends on the filter cutoff frequency and the proposed algorithm reduces the noise in the whole frequency band.

The proposed general algorithm can be seen in figure 3.4.

3.3 Experimental results applying the phase recovery algorithm

To verify the theoretical result obtained in the previous section, an experiment was carried out using as a base a signal composed of six harmonics of different amplitudes (0.4, 0.5, 0.6, 0.6, 0.3, 0.1), frequencies (50 Hz, 200 Hz, 400 Hz, 400 Hz, 600 Hz, and 700 Hz), and phases ($\pi/4$, $\pi/6$, $\pi/3$, $\pi/2$, $\pi/12$, and π), contaminated by white Gaussian noise of zero mean value and constant variance equal to one, generated with MATLAB. The sampling frequency used was 5000 Hz. Figures 3.5(a) and (b) show the contaminated signal and its spectrum, respectively. The results obtained by applying the proposed algorithm are shown in figure 3.5(c). As can be seen in figure 3.5(c), the high correlation levels compared with the useful (noise-free) signal are visually appreciable (see table 3.2).

Other types of periodic signals were also worked with in the simulation environment. In particular, a rectangular pulse train and a triangular pulse train were generated, which can be represented in Fourier series according to [16] and [17], respectively. Both the rectangular and triangular signals were generated with a frequency of 100 Hz. The data window used in both cases was 8192 samples. On the other hand, the sampling frequency used for the experiment with the rectangular

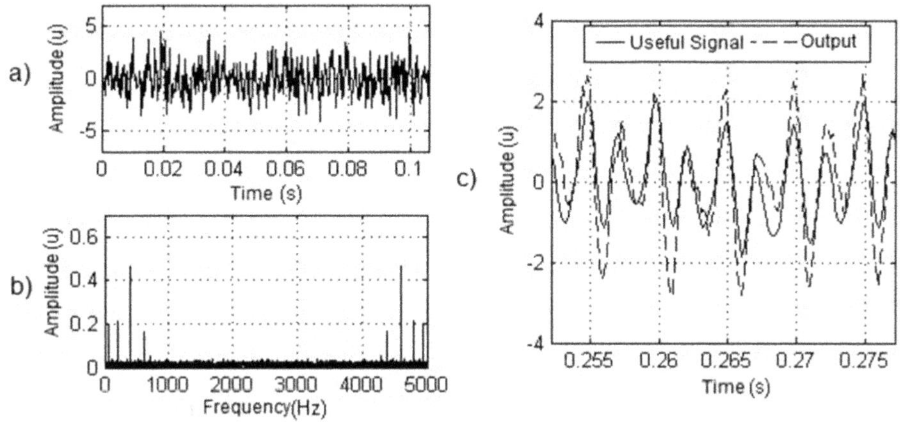

Figure 3.5. (a) Contaminated multitone signal. (b) Spectrum of the contaminated multitone signal. (c) Comparison between the useful signal (noise-free signal) and the output after evaluating the proposed algorithm.

Table 3.2. Results of the proposed noise reduction algorithm based on fourth-order statistics. Simulated signals. Average values obtained for 20 noise realizations.

Signal	Input SNR (dB)	Output SNR (dB)	Input correlation	Output correlation
Multitone	−3.93	5.28	0.543 7	0.760 1
Rectangular pulse train	−5.83	0.14	0.447 3	0.920 1
Triangular pulse train	−4.58	1.66	0.501 8	0.913 3

signal was 10 000 Hz, and the sampling frequency used for the triangular signal was 5000 Hz. Figures 3.6(a) and (b) show the contaminated rectangular pulse train signal and its spectrum, respectively. In addition, figure 3.6(c) shows a comparison between the useful (noise-free) signal and the output signal obtained from the processing after evaluating the proposed algorithm.

On the other hand, figures 3.7(a) and (b) show the contaminated triangular pulse train signal and its spectrum, respectively. In turn, figure 3.7(c) shows a comparison between the useful (noise-free) signal and the output signal obtained from the processing using the proposed algorithm.

Table 3.2 summarizes the results of the verification of the proposed algorithm for the simulated signals used in the experiments, the input parameters in relation to the SNR and correlation levels used, and those obtained after processing. These parameters (SNR and correlation) are a measure of the effectiveness of the proposed algorithm. The correlation index was used to evaluate the similarity between the ideal (noise-free) original signal and the output after evaluating the proposed algorithm.

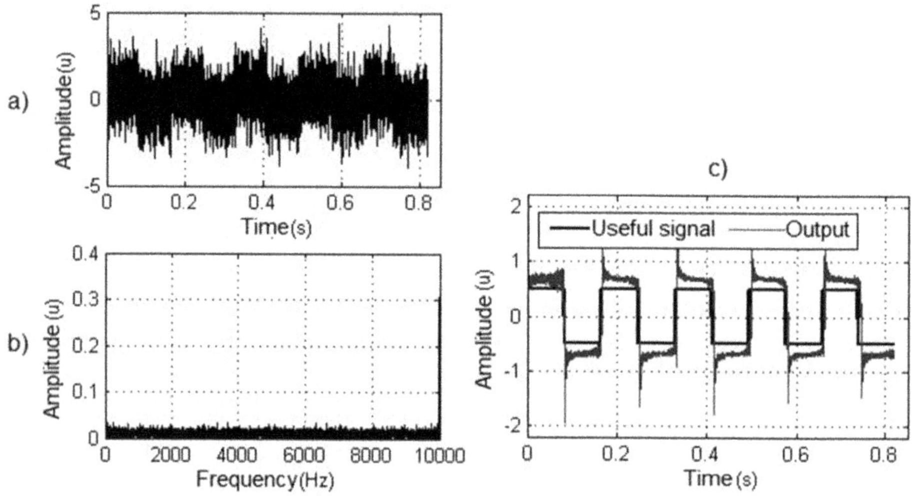

Figure 3.6. (a) Rectangular pulse train with noise. (b) Spectrum of the rectangular pulse train with noise. (c) Comparison between the useful signal (signal without noise) and the output after evaluating the proposed algorithm.

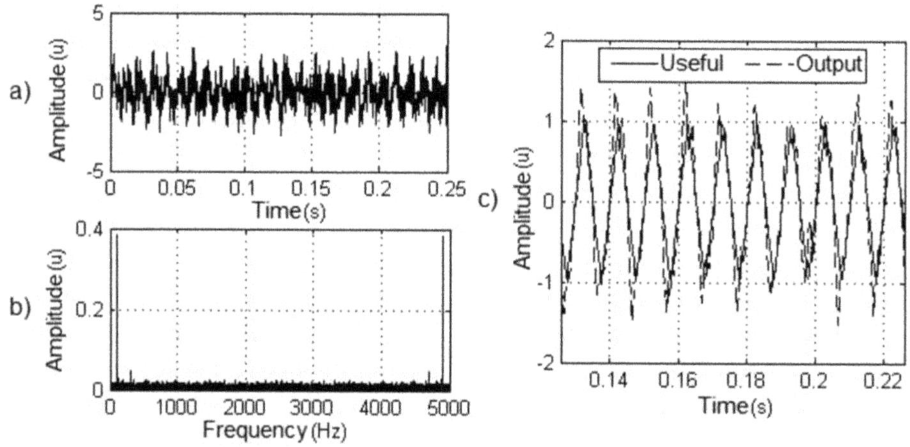

Figure 3.7. (a) Triangular pulse train with noise. (b) Spectrum of the triangular pulse train with noise. (c) Comparison between the useful signal (signal without noise) and the output after evaluating the proposed algorithm.

In all the experiments mentioned above we worked with the same noise level in terms of SNR values for different signal realizations by keeping the same length of the data window and taking an initial sample of 20 noise realizations, length 8192 samples, resulting in a data sample of 163 840 samples. The values shown in table 3.2 are the result of the average of the 20 processing operations according to the type of signal to be evaluated.

3.4 Computational cost analysis of the proposed method compared with others

As already mentioned in this work, the previous application of higher-order statistical analysis (HOSA) techniques to reduce noise presents the problem of the loss of the phase of the input signal. To avoid this, some methods have been developed based on higher-order spectra that combine one or several one-dimensional components of such spectra, taking advantage of the fact that the higher-order spectral features preserve the phase information. However, in such cases the computational cost, measured in execution time by number of samples to process, number of basic operations involved, and memory consumption, is too high: higher-order spectral features require the computation of multidimensional functions which does not make their practical implementation feasible.

To further illustrate the above statement, a comparative operational analysis is presented to evaluate the computational cost of some of the most relevant methods based on higher-order statistics, focused on the recovery of the phase information of a given signal.

It is necessary to highlight the sequential characteristic of the processing for each of the analyzed algorithms; likewise, the evaluation of the computational cost for each algorithm has focused on the analysis of the most critical blocks involved in the reconstruction of the phase information.

3.4.1 Computational cost of methods based on bispectrum computation

For a given real discrete signal $x(n)$ the bispectrum is defined as the Fourier transform of the third-order cumulant of the signal (see equation (2.14)). The steps performed for its calculation are the following. Let $x(n)$, $n = 1, \ldots, N$ be a random vector of length N, with mean value equal to zero. The first step is the calculation of the third-order cumulant sequence of the data vector $x(n)$. Then the second step is to calculate the third-order spectrum by means of the two-dimensional Fourier transform of the third-order cumulant sequence:

$$C_3^x(\omega_1, \omega_2) = \frac{1}{(2\pi)^2} \sum_{\tau_1=0}^{N-1} \sum_{\tau_2=0}^{N-1} c_3^x(\tau_1, \tau_2) e^{-j(\omega_1\tau_1+\omega_2\tau_2)}. \tag{3.13}$$

Once these calculations have been made, the phase of the sequence can be estimated.

Because the bispectrum preserves the phase information of each bispectral component. Then, the phase information can be obtained according to [18, 19], as follows:

$$\phi_{C_3^x}(\omega_1, \omega_2) = \varphi_x(\omega_1) + \varphi_x(\omega_2) - \varphi_x(\omega_1 + \omega_2), \tag{3.14}$$

where $\phi_{C_3^x}(\omega_1, \omega_2)$ is the phase matrix of each bispectral component. From equation (3.14) the phase vector of the original signal can be obtained by clearing $\varphi_x(\omega_1)$, using different methods [20, 21]. For example, taking equation (3.14), $\omega_1 = \omega$, and

$\omega_2 = \Delta\omega$, the limit is evaluated as $\lim_{\Delta\omega \to 0} \frac{\phi_{C_s^z}(\omega, \Delta\omega)}{\Delta\omega}$ as proposed by [20]. The idea is to obtain the corresponding phase vector from the output samples $\left(\phi_{C_3^s}(\omega_1, \omega_2)\right)$.

It can be observed that in the first step of the algorithm for the calculation of the bispectrum there are two multiplications and an accumulation for each sample to be processed, in addition to a normalization of each element, and three implicit cycles of the process, equivalent to the temporal indexes, resulting from the two-dimensional function of the third-order cumulant.

In step 2, the bispectrum of the signal is obtained through the two-dimensional Fourier transform whose algorithm makes its implementation even more complex [22].

This method is only valid for processes whose probabilistic distribution function is not symmetric, since for symmetrically distributed processes the third-order cumulant is zero, as mentioned previously in section 3.1 [14].

3.4.2 Computational cost of methods based on trispectrum computation

Some methods to reduce noise, applying HOSA, and also with the aim of preserving the phase, are based on the calculation of the trispectrum. The calculation of the trispectrum is based on the Fourier transform of the fourth-order cumulant (see equation (2.17)), whose computation is performed from the second- and fourth-order moments, which results in a three-dimensional function with a three-dimensional Fourier transform as well [22], making practical implementation inefficient.

Analogous to the bispectrum, the trispectrum preserves the phase information, since the phase of each trispectral component can be obtained by the relation given in [19], equal to

$$\phi_{C^f}(\omega_1, \omega_2, \omega_3) = \varphi_x(\omega_1) + \varphi_x(\omega_2) + \varphi_x(\omega_3) - \varphi_x(\omega_1 + \omega_2 + \omega_3), \tag{3.15}$$

where $\phi_{C_4^z}(\omega_1, \omega_2, \omega_3)$ is the phase matrix of each spectral component. Analogous to the method based on the bispectrum of the signal, algorithms such as those discussed in [23] are developed to obtain the phase information from equation (2.17); these are also based on recursive methods to recover the phase vector of the original signal from the output samples obtained in $\phi_{C_4^s}(\omega_1, \omega_2, \omega_3)$.

3.4.3 Computational cost of a method based on the combination of one- and

This method uses a combination of simultaneous regions, based on the frequency selection contents of one- and two-dimensional components of the cumulants. This method also has a computational cost that is much lower than those of methods based on the calculation of the trispectrum and relatively lower than that of the model based on the bispectrum, since it uses only one-dimensional and two-dimensional regions simultaneously and not all regions of the spectrum of higher order as the previous methods. On the other hand, this method has the disadvantage of being based on frequency selection criteria to find the best region for spectrum computation. Moreover, its computational cost is high since it involves two-dimensional functions [14]. The steps to follow for the algorithm based on this method are as follows:

1. Estimate the bispectrum of the sequence $x(n)$, $C_3^x(\omega_1, \omega_2)$.
2. Calculate the principal argument of the phases of two consecutive regions of the spectrum, for example, $C_3^x(\omega_1, i)$ and $C_3^x(\omega_1, i+1)$, where $i = 0, 1, 2, \dots, \omega_2$ and $0 \leqslant \omega_1 \leqslant N - 1$, where N is the length of the fast Fourier transform (FFT).
3. Estimate the phase of the sequence $x(n)$ based on equation (3.15).

3.4.4 Computational cost of the proposed phase recovery algorithm

To cover the shortcomings of the higher-order spectra–based methods is the task of reconstructing the phase information. As already mentioned, in this book a noise reduction algorithm is proposed that computes a higher-order feature, which is combined with the convolution process (see equation (3.9)). In detail, the convolution reflects a multiplication and an accumulation for each input sample, in addition to an incremental normalization. There is no need to perform spectral analysis to preserve phase information. The novelty of this procedure lies in the incorporation of additional information to the algorithm through the convolution process, since the previous methods do not have this characteristic, and in trying to obtain the phase information using only the output data. This procedure's computational complexity is far lower than those of the methods discussed so far in the current bibliographical references (see table 3.3), since its computation is completely one-dimensional and can be executed operationally at the same instant of time, according to its microelectronic implementation. Table 3.4 shows a comparative analysis of the number of operations involved in a general way in the groups of algorithms focused on phase recovery.

Table 3.3. Comparison of computational cost of statistical methods used for phase recovery.

Algorithm	Basic operations not including FFT	Cyclic operations	FFT
Bispectrum	Three multiplications Two accumulations	Three cycles	2D
Trispectrum	Four multiplications Three accumulations	Four cycles	2D
Combination of spectral regions	Two multiplications Two accumulations	Three or more	1D and/or 2D
Proposed phase recovery algorithm: convolution	One multiplication One accumulation	Two cycles Two cycles	No

Table 3.4. Comparison of the computational cost of the statistical methods used for phase retrieval with the proposed algorithm, as a function of the data window to be processed, using Windows O.S, RAM 2 GB, dual core processor.

Method	Minimum SNR value	Reference
Algorithms based on HOSA	0 dB, −11.5 dB	[24, 25]
	−14 dB	Proposed algorithm

Table 3.5. Comparison of the computational cost of the statistical methods used for phase retrieval, with the proposed algorithm, for a maximum window of 64 samples in MATLAB, using Windows O.S, RAM 2 GB, dual core processor.

Algorithm	Runtime in seconds
Bispectrum	0.075 9
Trispectrum	0.206 0
Combination of spectral regions	0.055 8
Proposed phase recovery algorithm: convolution	1.851 2e − 4
Proposed algorithm for phase retrieval noise reduction	0.002 7

To obtain a validation of the execution time of the algorithms described in table 3.5, a simulation analysis of each of them was carried out using MATLAB, and their execution time was obtained for a maximum sample window of length 64. The execution time of each of the algorithms was obtained through the *tic-toc* function of MATLAB, which measures the computation time in seconds of a given function.

For the evaluation of the runtime analysis, ten runtime measurements were taken for each algorithm per observation of processed samples, and the average of the ten measurements taken in each processed data window is issued as the resulting time value.

The results obtained can be seen in figure 3.8, which shows an excessive time consumption for each of the functions that use higher-order spectra, in contrast with the behavior of the algorithm of recovery of phase information proposed in this work.

It should be mentioned that all the HOSA functions used to program the algorithms included in table 3.5, with the exception of the proposed algorithm, are part of the HOSA – Higher Order Spectral Analysis Toolbox for MATLAB [7].

It should also be noted that the amplitude distorted during the calculation process for the recovery of the phase information by all the analyzed algorithms is not taken into account, with the exception of the proposed algorithm for noise reduction and phase recovery (figure 2.4). The execution time given for the proposed algorithm is the time necessary to obtain at the output an approximate estimate of the useful signal free of noise.

Table 3.5 shows as additional data the total execution time of the HOSA-based algorithms for the maximum window length, in this case 64 samples, including the

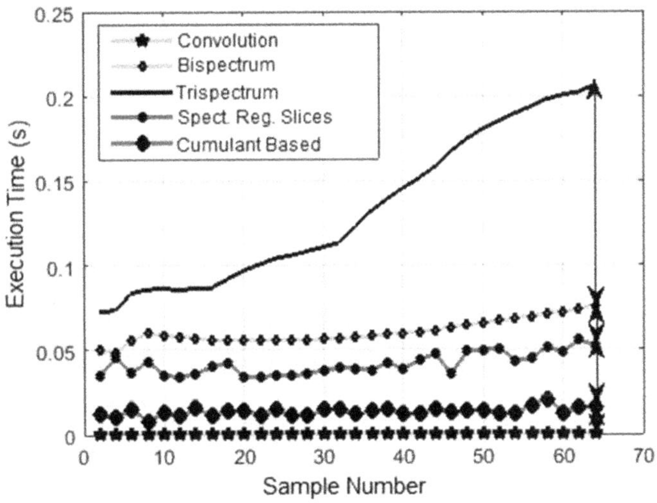

Figure 3.8. Comparison of the computational cost of the statistical methods used for phase retrieval with the proposed algorithm, as a function of the data window to be processed, using Windows O.S, RAM 2 GB, dual core processor.

proposed algorithm for phase information retrieval. The execution time of the proposed algorithm for phase retrieval is 301.4 times faster than the fastest of the other algorithms using spectral region combining. Furthermore, the execution time of the proposed algorithm for noise reduction and phase retrieval (figure 2.4) is shown as an additional data point which is still 20.67 times faster than the algorithm based on the combination of spectral regions.

It is necessary to emphasize that the time data shown so far are a relative measure of the execution time for both algorithms taking a given data window. The calculated execution time, although it is measured only during the processing of the algorithms in each case, depends on the processes of the operating system in execution, interruptions to the processor, and other factors that can progressively influence the calculation. Nevertheless, it gives an idea of the efficiency of the proposed algorithm with respect to the proposals currently referred to, based on HOSA.

From the results seen in the comparative analysis, it can be concluded that the practical implementation of the proposed algorithm for phase recovery, based on a convolution process, is much more efficient than other methods based on higher-order spectra, since it solves the problem of loss of phase information while requiring for its implementation a lower computational cost. Likewise, the proposed algorithm reduces noise and phase recovery.

3.5 SNR levels processed by the proposed algorithm compared with others developed for noise reduction and phase retrieval

To check the effectiveness of the proposed algorithm based on the combination of the one-dimensional component of the fourth-order cumulant and a convolution process, table 3.6 shows the minimum SNR levels used in the works referred in the

Table 3.6. Comparison of the SNR values used by the proposed cumulant-based algorithm with those used by previous works consulted in the literature.

Method	Minimum SNR	Reference
HOSA algorithms	0 dB, −6 dB, 12.5 dB, −20 dB, 3 dB, −10 dB, 5 dB	[6, 7, 18, 26–30]
Proposed algorithm	−3.93 dB, −5.83 dB, −6.09 dB	Multitone, square, and triangular signal

literature based on HOSA for a given application and compares them with the levels obtained by the proposed algorithm.

It is necessary to emphasize that none of the works consulted evaluated the levels of correlation existing at the output in comparison with the waveform obtained after processing, nor did they compare the waveform obtained as a form of visual verification of the effectiveness of the algorithm, which is done in this research.

In the references where work is done with a lower SNR than the one used in the experiments, for example in [2] where −20 dB is used to test the experiments, this work has to do with the detection of the amplitude, frequency, and phase components in harmonic signals.

The SNR used in reference [31] is related to the use of HOSA for digital modulation classification, and in the work done in [4] it is based on reducing noise in cryptographic channels.

In the previous works, the waveform at the output is not obtained, and it is not compared with the input signal in terms of correlation levels, since that is not the objective of these works.

Tables 3.7, 3.8, and 3.9 are shown to check the SNR levels with which the algorithm can work, to obtain correlation values above 0.45. Tables 3.7, 3.8, and 3.9 show the SNR levels at the input and output as well as the correlation levels obtained at the output after processing. The results shown refer to the multitone, rectangular, and triangular signals used in section 3.3. The experiments were carried out until obtaining maximum correlation values at the output close to or greater than 0.5. In all the experiments performed, we worked with the same characteristics of the noise in terms of zero mean value and constant variance equal to one, decreasing the SNR values for different realizations of the signal, keeping the same length of the data window, and taking an initial sample of 20 realizations of noise of length 8192 samples, analogous to the experiment of section 3.3. The values shown in all the tables are the result of the average of the 20 processing operations according to the type of signal to be evaluated.

Another analysis was also carried out to check the minimum SNR levels. In this case, an experiment was performed, taking as useful signal a cosine of amplitude 0.1 and frequency 100 Hz, sampled at 2 kHz and taking a window length of 1024 samples. This experiment was carried out to obtain the minimum SNR levels for which a harmonic signal can be detected, taking into account that a correlation between input and output of at least 0.5 is obtained. Figure 3.9 shows the SNR

Table 3.7 Results obtained for the multitone signal used, decreasing the SNR values at the input. Average values obtained for 20 noise realizations.

Signal	SNR input	SNR output	Correlation input	Correlation output
Multitone	−3.937 3	5.284 5	0.543 7	0.760 1
	−4.497 9	5.217 7	0.519 5	0.749 0
	−5.024 5	5.152 1	0.497 1	0.736 6
	−5.520 9	5.084 8	0.476 3	0.723 0
	−5.990 5	5.014 2	0.457 0	0.708 4
	−6.436 1	4.946 4	0.439 0	0.692 9
	−6.859 9	4.849 8	0.422 3	0.677 0
	−7.263 9	4.816 3	0.406 7	0.661 0
	−7.650 0	4.756 5	0.392 2	0.644 8
	−8.019 7	4.701 7	0.378 6	0.628 3
	−8.374 3	4.651 5	0.365 9	0.611 6
	−8.714 9	4.606 9	0.353 9	0.594 5
	−9.042 8	4.569 6	0.342 7	0.577 4
	−9.358 6	4.538 3	0.332 2	0.560 1
	−9.663 4	4.512 4	0.322 3	0.542 9
	−9.957 9	4.491 8	0.312 9	0.525 9

Table 3.8 Results obtained for the square signal used, decreasing SNR values at the input. Average values obtained for 20 noise realizations.

Signal	SNR input	SNR output	Correlation input	Correlation output
Rectangular pulse train	−5.837 8	0.149 1	0.447 3	0.920 1
	−6.665 7	0.197 0	0.413 2	0.902 3
	−7.421 4	0.251 8	0.383 3	0.878 9
	−8.116 7	0.290 6	0.357 1	0.849 3
	−8.760 4	0.340 0	0.333 9	0.812 8
	−9.359 6	0.373 0	0.313 3	0.768 4
	−9.920 2	0.396 7	0.294 9	0.717 0
	−10.446 8	0.420 6	0.278 4	0.660 9
	−10.943 2	0.449 4	0.263 5	0.601 5
	−11.412 9	0.471 3	0.250 1	0.541 3
	−11.858 4	0.486 5	0.237 8	0.483 7

values obtained at the output as a function of the input SNR values. Likewise, figure 3.10 shows the correlation values obtained at the output as a function of those referred to the input when the algorithm based on cumulants is applied to a harmonic signal contaminated by noise. The values represented in both figures are the average of 20 noise realizations taken as a sample, in the experiment, which is equivalent to a noise data window of 20 480 samples.

Table 3.9 Results obtained for the triangular signal used, decreasing the SNR values at the input. Average values obtained for 20 noise realizations.

Signal	SNR input	SNR output	Correlation input	Correlation output
Triangular	−4.586 7	1.664 9	0.501 8	0.913 3
	−5.414 5	1.645 3	0.465 9	0.880 2
	−6.170 3	1.626 8	0.434 1	0.840 5
	−6.865 5	1.612 8	0.405 8	0.795 4
	−7.509 2	1.607 3	0.380 7	0.746 9
	−8.108 5	1.612 7	0.358 1	0.696 7
	−8.669 1	1.630 7	0.337 9	0.646 4
	−9.195 6	1.661 5	0.319 6	0.597 4
	−9.692 1	1.702 1	0.303 1	0.550 4
	−10.161 7	1.750 3	0.288 0	0.505 9
	−10.607 3	1.812 0	0.274 3	0.464 4

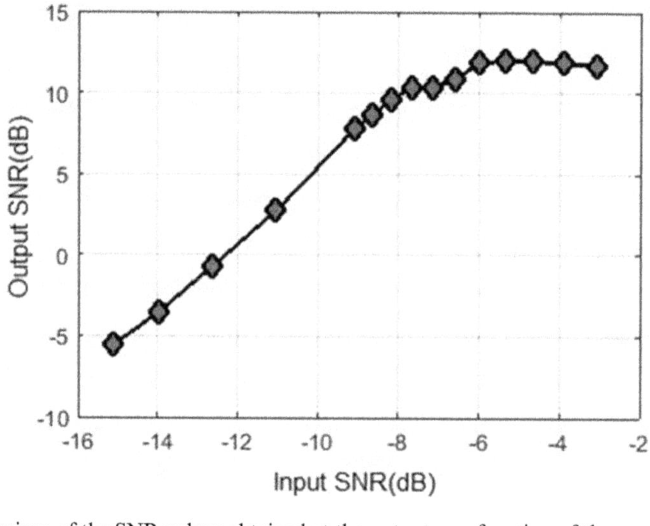

Figure 3.9. Comparison of the SNR values obtained at the output as a function of those referred to the input, for a cosine signal. Mean values obtained for 20 noise realizations.

In general, it can be said that the noise levels with which the proposed algorithm based on the combination of higher-order statistics and convolution can work depend on the characteristics of the noise process, the signal, the work environment where the processing is performed, and the particular application where its use is required. In addition, according to the results obtained, it can be said that the effectiveness of the proposed algorithm and its performance are similar or superior to the works discussed in the literature and cited in table 3.6, taking into account the SNR levels with which it works in previous works consulted. In addition, the proposed algorithm has a lower computational cost compared with the works carried out and discussed in the literature.

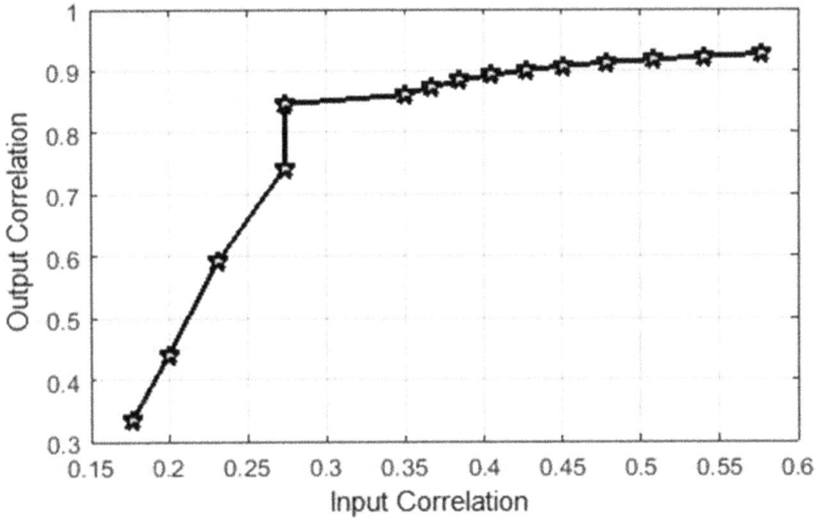

Figure 3.10. Comparison of the correlation values obtained at the output as a function of those given at the input, for a cosine signal. Average values obtained for 20 noise realizations.

3.6 Comparative analysis according to other noise reduction methods not based on HOSA

In order to compare the result of the proposed algorithm with other algorithms not based on HOSA, an experiment was carried out for different SNR and correlation levels at the input. The experiment used a multitone signal consisting of nine harmonics of different amplitude (0.1, 0.2, 0.3, 0.5, 1.0, 0.8, 0.6, 0.4, 0.9), frequency (50 Hz, 200 Hz, 300 Hz, 100 Hz, 500 Hz, 150 Hz, 600 Hz, 350 Hz, and 250 Hz), and phase in radians ($\pi/4$, $\pi/3$, $\pi/6$, $\pi/8$, $\pi\pi/2$, $\pi/5$, $3\pi/4$, and $2\pi/3$), contaminated by Gaussian white noise of zero mean value and constant variance equal to 1, generated through MATLAB. In the experiments we used 20 noise realizations, which were used to statistically measure the values obtained at the output referred to correlation and SNR. The sampling frequency used was 10 kHz. The data window for processing was 8192 samples.

For the comparison, an adaptive noise reduction scheme was used in a one-input one-output system, obtaining the reference signal for the adaptation of the filter coefficients from the contaminated input signal. The results obtained by the proposed algorithm were compared with those obtained by the least mean square (LMS) algorithm and the recursive least square (RLS) algorithm. To generate the noise correlated with the interfering signal we used as reference the output of a moving average filter of order one which has as input the contaminated multitone signal sample.

The adaptive noise reduction scheme used is similar to the one illustrated in figure 2.4. With this scheme a correlation between the noise of the contaminated sample and the noise obtained to be used as reference signal of 0.707 6 was obtained. The obtained results are shown in tables 3.10, 3.11, and 3.12, respectively, where the variation of the SNR values at the input using the proposed algorithm and the

Table 3.10 Results obtained by varying the SNR values at the input using the proposed algorithm. Average values obtained for 20 noise realizations.

Signal	Input SNR	Input correlation	Output SNR	Output correlation
	−7.698 6	0.383 2	0.014 8	0.757 9
Multitone	−10.054 2	0.302 1	0.014 6	0.640 8
with	−11.905 3	0.248 6	0.013 6	0.524 8
proposed	−13.430 4	0.210 9	0.013 0	0.451 6
algorithm	−14.727 2	0.183 0	0.012 6	0.369 2
	−15.855 4	0.161 7	0.012 3	0.286 7
	−16.853 7	0.144 8	0.012 2	0.216 8
	−17.749 0	0.131 1	0.012 2	0.169 4

Table 3.11 Results obtained by varying the SNR values at the input using the LMS adaptive filtering algorithm. Filter order of 5 and convergence factor 0.001. Mean values obtained for 20 noise realizations.

Signal	Input SNR	Input correlation	Output SNR	Output correlation
	−7.698 6	0.383 2	−1.188 6	0.638 3
Multitone	−10.054 2	0.302 1	−3.642 2	0.525 0
with LMS	−11.905 3	0.248 6	−5.925 3	0.422 6
algorithm	−13.430 4	0.210 9	−8.407 5	0.325 5
	−14.727 2	0.183 0	−12.507 8	0.205 6
	−15.855 4	0.161 7	−21.338 1	0.072 1
	−16.853 7	0.144 8	−33.948 4	0.014 3
	−17.749 0	0.131 1	−51.964 7	0.002 9

Table 3.12 Results obtained by varying the SNR values at the input using the RLS adaptive filtering algorithm. Filter order of 5 and forgetting factor 0.996 1. Average values obtained for 20 noise realizations.

Signal	Input SNR	Input correlation	Output SNR	Output correlation
Multitone	−7.698 6	0.383 2	−0.057 2	0.705 7
	−10.054 2	0.302 1	−2.398 6	0.605 6
	−11.905 3	0.248 6	−4.242 3	0.524 5
	−13.430 4	0.210 9	−5.777 0	0.459 2
	−14.727 2	0.183 0	−9.484 7	0.324 0
	−15.855 4	0.161 7	−13.512 1	0.212 3
	−16.853 7	0.144 8	−9.288 5	0.324 6
	−17.749 0	0.131 1	−10.371 0	0.287 7

adaptive methods are shown, as well as the results obtained in each case, for the multitone signal used.

Figures 3.11 and 3.12 show graphically a comparison of the SNR and correlation values obtained at the output as a function of the input values, by the proposed algorithm, compared with those obtained by the adaptive filtering methods used.

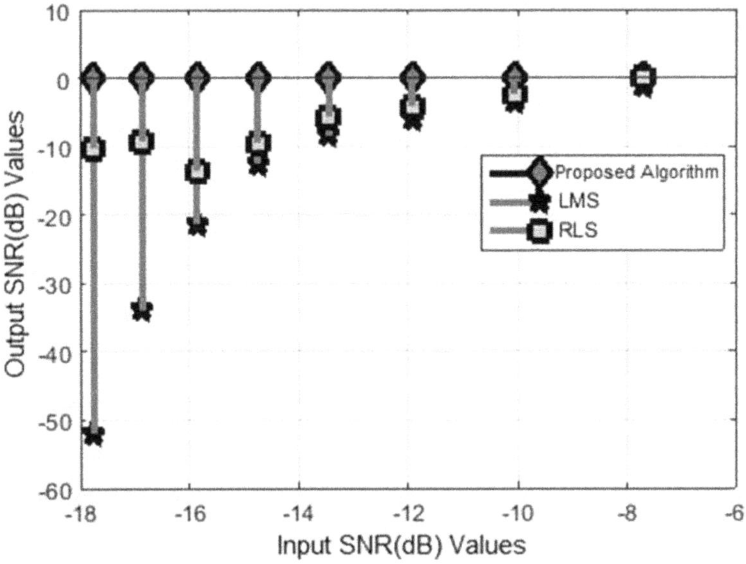

Figure 3.11. Comparison of the results obtained by the proposed algorithm with the LMS and RLS adaptive filtering algorithms, varying the SNR values at the input. Average values obtained for 20 noise realizations.

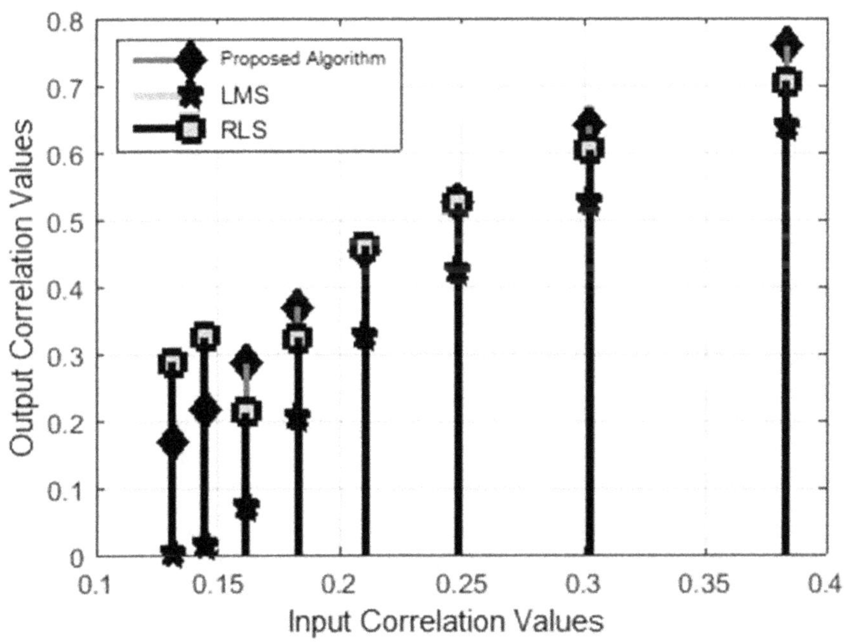

Figure 3.12. Comparison of the results obtained by the proposed algorithm with the LMS and RLS adaptive filtering algorithms, varying the correlation values at the input. Average values obtained for 20 noise realizations.

3.7 Application to noise reduction in real signals

In this section we will discuss the application of high-order statistical analysis for use on real signals such as vibrations, digital modulations, and the human tremor signal. In each application, the characteristics of the experiment and the types of signals in each case are shown, as well as the results obtained from the point of view of correlation levels and SNR values at the input and output of the processing.

3.7.1 Vibration sensor ADXL203

For experimental validation on real signals, in a first case we worked with the signal coming from a vibration sensor, ADXL203 of Analog Devices Inc. [17], digitized by means of a data acquisition system constituted by the Pmod AD1 analog-to-digital converter (ADC) module of the company Digilent Inc. [32]. This module uses one of the two AD7476A ADCs that are internally incorporated [33].

The Pmod AD1 has internally two low pass filters (one for each conversion channel) of 500 kHz bandwidth, of which only one is used to filter the signal before it enters the ADC in order to avoid the so-called aliasing effect [33].

The analog signal to be converted is in the range of 0 V to 3.3 V, the latter being the reference voltage of the converter used, whose resolution is 12 bits and whose maximum sampling frequency is 1 MHz; the latter is the frequency at which the experiment was performed.

The digitization process was controlled by an FPGA of the Spartan3E family [34], with which the digitization system was designed. In a first experiment, this sensor was used to measure human tremor, since this type of movement can be considered as a rhythmic oscillation (periodic signal) [35–37].

Figure 3.13 shows the spectrum of the contaminated ADXL203 sensor signal, and figure 3.14 shows a comparison between the signal obtained after evaluating the proposed algorithm and the original sensor signal.

The work with the signal obtained from the accelerometer (ADXL203) consisted of adding real noise from the ADXL203 sensor when there was no vibration activity.

When the ADXL203 accelerometer signal was digitized, it was found that the contaminating noise in this signal presented characteristics of a Gaussian nature. Figure 3.15 shows the sensor signal contaminated by real sensor noise obtained when there is no vibration activity (stationary sensor). Figure 3.16 also shows the histogram of the sensor noise, using the MATLAB *histfit* function.

As can be seen in figure 3.14, the output of the proposed algorithm and the original sensor signal show high levels of correlation, which demonstrates the effectiveness of the proposed algorithm. Table 3.13 summarizes the results obtained by applying the algorithm based on higher-order statistics for the ADXL203 vibration sensor signal, as well as the input parameters in relation to the SNR and correlation levels used and those obtained after processing.

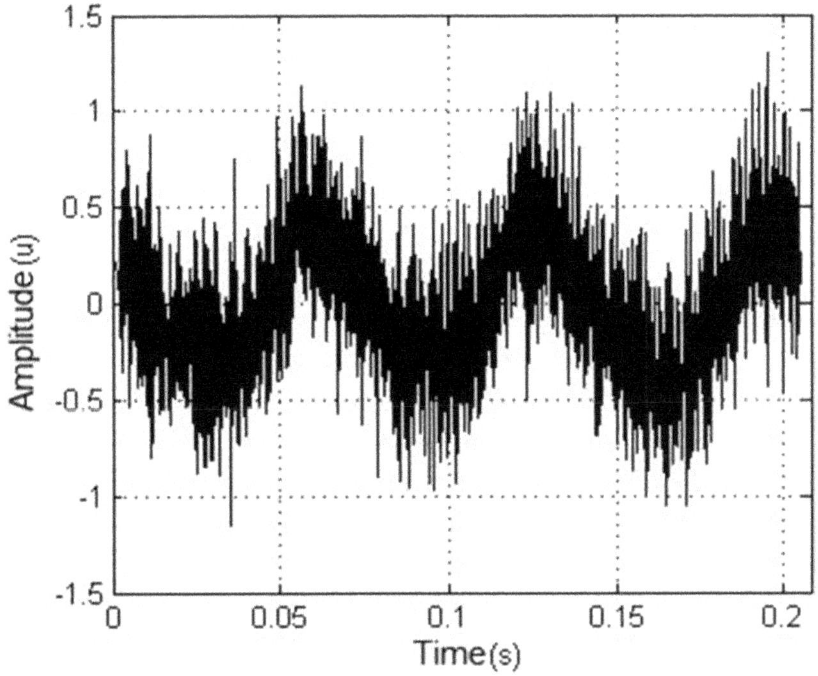

Figure 3.13. Sensor signal spectrum.

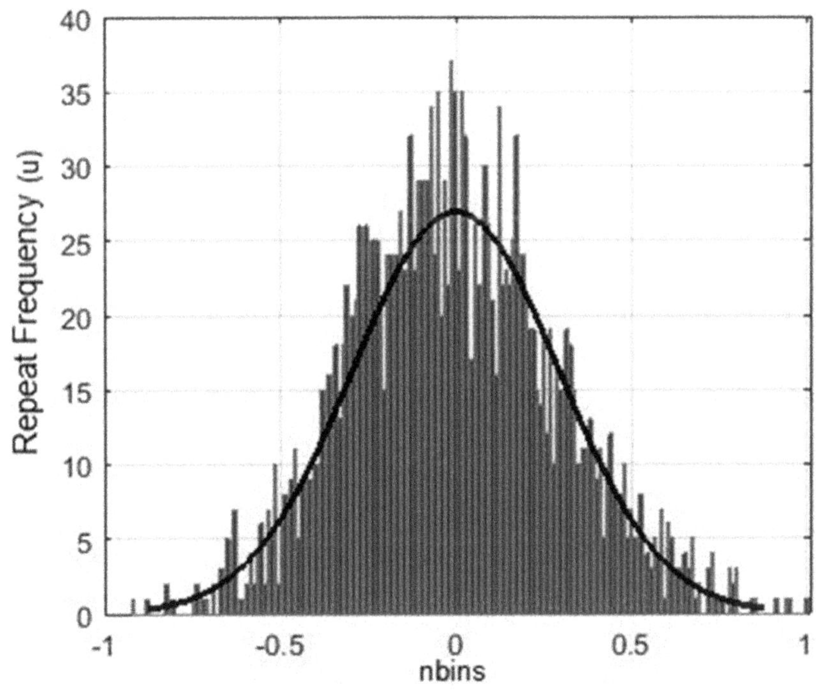

Figure 3.14. Comparison between the useful signal (sensor without noise) and the output after evaluating the proposed algorithm.

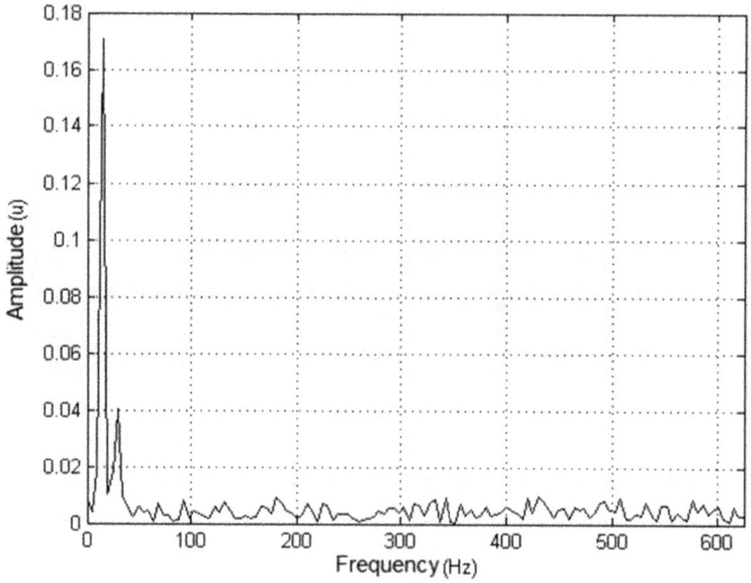

Figure 3.15. ADXL203 sensor signal with noise.

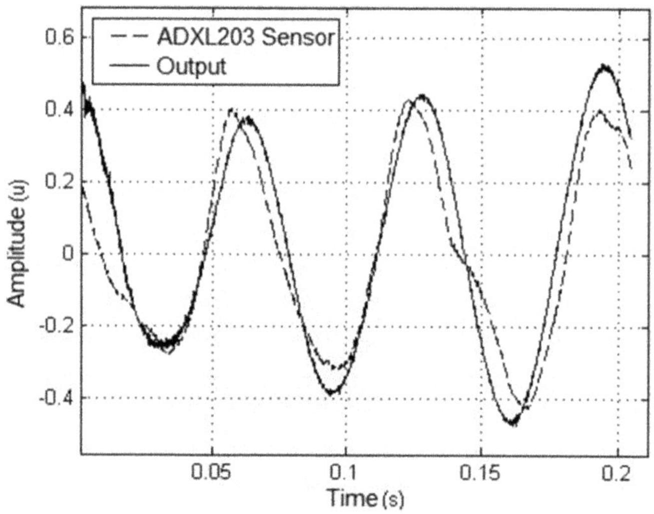

Figure 3.16. Histogram of sensor noise, using the MATLAB *histfit* function.

3.7.2 Application to noise reduction in digital modulations

The application of the proposed algorithms to the digital modulations used twenty samples of real modulations of the multiple phase shift keying type.

From the actual processed samples only the carrier frequency is known (see table 3.14).

Table 3.13 Results of the proposed noise reduction algorithm based on fourth-order statistics. ADXL203 sensor signal.

Signal	Input SNR (dB)	Output SNR (dB)	Input correlation	Output correlation
ADXL203 sensor	−1.291 1	7.998 9	0.654 7	0.932 9

Table 3.14 Application of the proposed algorithms to digital modulations contaminated by noise.

Signal	Estimated input SNR (dB)	Estimated output SNR (dB)	Amplitude peak (u)	Carrier frequency (Hz)
1_2PSK.wav	−6.331 2	−0.026 5	0.32	206.42
2_2PSK.wav	−10.257 1	−1.589 2	0.80	206.42
3_2PSK.wav	−6.692 5	1.466 2	0.35	206.42
4_2PSK.wav	−0.190 2	4.007 3	0.25	206.42
5_2PSK.wav	−14.356 9	−5.657 6	0.42	206.42
6_2PSK.wav	−9.123 9	0.294 8	0.30	206.42
7_2PSK.wav	−7.425 1	0.640 7	0.26	206.42
8_2PSK.wav	−1.130 4	8.589 4	0.06	2083
9_2PSK.wav	0.080 4	7.654 1	0.10	2083
10_2PSK.wav	−2.204 6	5.450 6	0.05	2083
11_4PSK.wav	−2.582 9	5.041 2	0.12	1964
12_4PSK.wav	−9.200 3	−2.823 9	0.62	1512
13_4PSK.wav	−1.336 2	6.019 1	0.54	2160
14_4PSK.wav	−6.681 6	0.233 1	0.60	1512
15_4PSK.wav	−4.121 4	−0.074 4	0.64	880
16_8PSK.wav	−10.695 9	−0.588 3	0.21	1500
17_8PSK.wav	−10.665 1	−3.363 0	0.40	1800
18_8PSK.wav	−13.802 2	−17.516 2	0.31	1800
19_8PSK.wav	−20.191 1	−15.041 8	0.75	10 500
20_8PSK.wav	−19.705 0	−18.831 5	0.51	1800

Both the input SNR and the output SNR were estimated, since only the contaminated input signal was available. For the SNR estimation, the Akaike algorithm [38] was used, which is an estimation method based on an information criterion (the Akaike information criterion, or AIC, is an estimator of prediction error and thereby relative quality of statistical models for a given set of data) that provides a measure of quality or accuracy according to a given situation. It tries to find the most accurate model for a given data set [38–40]. It is based on the power spectrum of the signal, for which it assumes that a certain number of spectral components taken as eigenvalues correspond to the power of the useful signal; the rest of the spectral components correspond to noise. Equation (3.16) is used to estimate the number of spectral components that make up the useful signal power:

$$AIC(k) = (N - k)L\log(\alpha(k)) + k(2N - k)$$
$$\alpha(k) = \left(\frac{1}{N - k}\right)\frac{\left(\sum_{i=k+1}^{N}\lambda_i\right)}{\left(\prod_{i=k+1}^{N}\lambda_i\right)^{1/(N-k)}}, \tag{3.16}$$

where N represents the number of points of the FFT, k is the assumed number of values corresponding to the signal power, and L is the number of processed samples. The spectral components of the power spectrum of the contaminated signal are sorted in descending order (from highest to lowest) up to the minimum value estimated by the algorithm, which indicates the position in the data vector corresponding to the values of the spectral components of the power spectrum containing the useful signal; the rest of the spectral components, from the obtained position onwards, correspond to the power of the noise [40, 41]. The obtained results, shown in table 3.14, reveal an increase of the SNR at the output compared with the estimated SNR at the input.

It is valid to emphasize that the contaminating noise in each of the processed modulations is present in the useful signal band, which makes the processing more complex, from the point of view that the algorithm should only be able to reduce the noise and not eliminate harmonics corresponding to information in the useful signal. Figure 3.17 shows a comparison between the input contaminated signal (in black) and the output obtained after evaluating the proposed algorithm (in grey).

In the processed modulations the noise is present in the useful signal band, which causes the amplitudes of the harmonics containing information to be affected by the incident noise in each given frequency component, which is resolved once the signal has been processed.

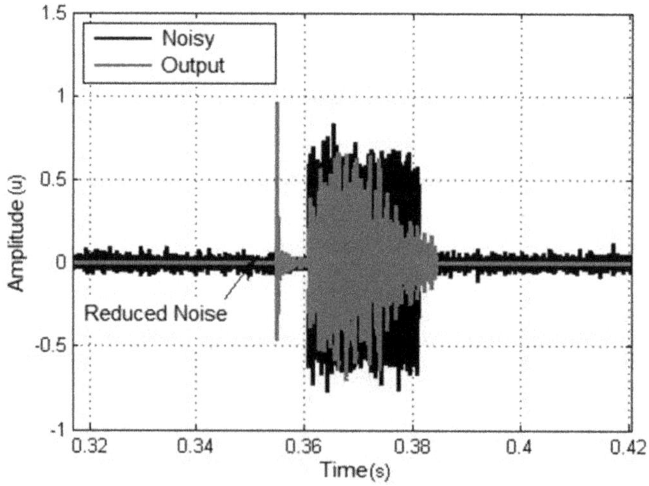

Figure 3.17. Input–output comparison signal 19_8PSK.wav.

3.7.3 Noise reduction in the human tremor signal

In the experiment applying the algorithms to the human tremor signal, real samples (taken from the measurements of six patients with Parkinson's disease in the office of the Provincial Hospital Abel Santamaría Cuadrado of Pinar del Rio, Cuba) were used as part of the Institutional Project of the University of Pinar del Rio: 'Quantitative Measurement of the Human Tremor'. Several works have been carried out so far in relation to the signal processing of human tremor [42–48], focusing on the characterization and measurement of tremor, including noise reduction given by involuntary movements or other factors. In Parkinson's patients at rest, the frequency of the fundamental harmonic is around 3 Hz to 8 Hz. The sampling frequency used in the measurements was 142.9 Hz, and the samples were digitized through the system designed in [44].

Table 3.15 shows the results obtained from the measurements, which were taken with the patients at rest and in a posture with their arms extended in tension. The estimated SNR values were obtained using the Akaike algorithm, similar to that used for digital modulations. As can be seen, the SNR levels estimated at the output of the proposed algorithm are higher compared with the levels estimated at the input.

The proposed algorithm is based on the statistical characteristics of the additive model of useful signal plus contaminating signal, discriminating the interfering random process from the useful information to be processed, so that the noise in the useful frequency band is reduced.

Unlike other procedures, the algorithm treated in this research does not depend on any information of the contaminated input signal, only on the periodicity parameter in the sample to be processed and the stationarity of the interfering random process.

3.8 Conclusions of the chapter

In this chapter the problem of applying HOSA to reduce Gaussian noise in periodic signals was demonstrated theoretically and experimentally, given the loss of the phase information after the application of the higher-order statistic. The high

Table 3.15 Results of the application of the algorithm based on the fourth-order statistics noise reduction in the human tremor signal.

Signal	Est. input SNR (dB)	Est. output SNR (dB)	Amp. peak spectrum (u)	Measurement characteristics
Patient_1	−3.572 1	0.237 2	0.14	At rest
Patient_2	−17.846 1	−4.449 3	0.002	At rest
Patient_3	−7.934 6	1.044 9	0.013	In tension
Patient_4	−3.248 0	4.585 0	0.17	At rest
Patient_5	−16.835 1	−4.528 9	0.001 51	In tension
Patient_6	−15.879 3	−2.603 4	0.001 67	In tension

computational cost of the higher-order statistics that try to preserve phase information was also confirmed from the operational point of view and in terms of execution time.

Furthermore, the theoretical and practical foundations that support the proposal of a new algorithm for obtaining phase information, based on a convolution process, were discussed throughout the chapter.

A comparative analysis was presented from the quantitative point of view in terms of computational costs, with respect to the existing algorithms based on HOSA, oriented to noise reduction while preserving the phase information; in this way, the feasibility of implementation of the proposal presented was demonstrated.

Likewise, the verification of the proposed algorithm based on joint techniques of statistical signal processing, convolution, and spectral estimation, from the experimental point of view, was carried out through MATLAB, using both simulated signals, proving the effectiveness of the algorithm itself through the correlation indexes, and SNR obtained at the output in comparison with the input data. It was demonstrated that the proposed algorithm is not only more effective in terms of SNR than previous similar techniques but also of lower computational cost. An improvement in computational cost was obtained with respect to the existing methods, in relation to the number of operations involved and obtaining a ratio of 301.4 times higher computational speed with respect to the other algorithms based on HOSA for phase recovery. In addition, a reduction ratio of the execution time of 20.67 was obtained, including the whole process of noise reduction and phase recovery.

In addition in this chapter an analysis was made to evaluate the effectiveness of the algorithm according to the length of the data window to process, the number of samples per period, the number of periods of the signal in the data window, and the variation of the sampling frequency used. Furthermore, applications to the reduction of noise in vibration signals, digital modulations, and the treatment of noise in biosignals such as human tremor were treated, obtaining satisfactory results in all cases, taking into account the SNR levels obtained at the output in comparison with the levels presented at the input before processing.

References

[1] Swami A and Mendel J M 1988 Cumulant-based approach to the harmonic retrieval problem *ICASSP-88, Int. Conf. on Acoustics, Speech, and Signal Processing* **4** 2264–7

[2] Swami A and Mendel J M 1991 Cumulant-based approach to harmonic retrieval and related problems *IEEE Trans. Signal Process.* **39** 1099–109

[3] Le T-H, Clediere J, Serviere C and Lacoume J-L 2007 Noise reduction in side channel attack using fourth-order cumulant *IEEE Trans. Inf. Forensics Secur.* **2** 710–20

[4] Zhang Y and Wang S-X 1998 A hybrid approach to harmonic retrieval in non-Gaussian noise using fourth-order moment and autocorrelation *ICSP '98 4th Int. Conf. on Signal Processing* **1** 411–4

[5] Blagouchine I V and Moreau E 2009 Unbiased adaptive estimations of the fourth-order cumulant for real random zero-mean signal *IEEE Trans. Signal Process.* **57** 3330–46

[6] Iglesias Martínez M E and Hernández Montero F E 2013 Detection of periodic signals in noise based on higher-order statistics joined to convolution process and spectral analysis *Progress in Pattern Recognition, Image Analysis, Computer Vision, and Applications* ed J Ruiz-Shulcloper and G Sanniti di Baja (Berlin: Springer) pp 488–95

[7] Swami A, Mendel J M and Nikias C L 2001 HOSA—Higher Order Spectral Analysis Toolbox for Use with MATLAB. MATLAB Central File Exchange http://www.mathworks.com/matlabcentral/fileexchange/3013-hosa-higher-order-spectral-analysis-toolbox

[8] Salavedra Molí J M 1995 Técnicas de speech enhancement considerando estadísticas de orden superior *Doctoral Thesis* Universitat Politecnica de Catalunya 1995 http://hdl.handle.net/10803/6943

[9] Vaseghi S V 2008 *Advanced Digital Signal Processing and Noise Reduction* 4th edn (New York: Wiley)

[10] Nikias C L and Mendel J M 1993 Signal processing with higher-order spectra *IEEE Signal Process. Mag.* **10** 10–37

[11] Rabiner L R, McClellan J H and Parks T W 1975 FIR digital filter design techniques using weighted Chebyshev approximation *Proc. IEEE* **63** 595–610

[12] Oppenheim A V and Schafer R W 1989 *Discrete-Time Signal Processing* (Englewood Cliffs, NJ: Prentice-Hall)

[13] Petropulu A P and Nikias C L 1992 Signal reconstruction from the phase of bispectrum *IEEE Trans. Signal Process.* **40** 601–10

[14] Petropulu A P and Pozidis H 1998 Phase reconstruction from bispectrum slices *IEEE Trans. Signal Process.* **46** 527–30

[15] Weisstein E W Fourier series—square wave. MathWorld—A Wolfram Web Resource http://mathworld.wolfram.com/FourierSeriesSquareWave.html

[16] Weisstein E W Fourier series—triangle wave. MathWorld—A Wolfram Web Resource http://mathworld.wolfram.com/FourierSeriesTriangleWave.html

[17] ADXL203 Accelerometer Data Sheet Rev E. Analog Devices http://www.analog.com/en/mems-sensors/mems-inertial-sensors/adxl203/products/product.html

[18] Anderson J M M, Giannakis G B and Swami A 1995 Harmonic retrieval using higher order statistics: a deterministic formulation *IEEE Trans. Signal Process.* **43** 1880–9

[19] Giannakis G B and Mendel J M 1989 Identification of nonminimum phase systems using higher order statistics *IEEE Trans. Acoust. Speech Signal Process.* **37** 360–77

[20] Matsuoka T and Ulrych T J 1984 Phase estimation using the bispectrum *Proc. IEEE* **72** 1403–11

[21] Holambe R S, Ray A K and Basu T K 1996 Signal phase recovery using the bispectrum *Signal Process.* **55** 321–37

[22] Geng M, Liang H and Wang J 2011 Research on methods of higher-order statistics for phase difference detection and frequency estimation *4th Int. Congress on Image and Signal Processing* **4** 2189–93

[23] Dianat S A and Raghuveer M R 1990 Fast algorithms for bispectral reconstruction of two-dimensional signals *Int. Conf. Acoustics, Speech, and Signal Processing* pp 2377–9

[24] Yan X and Jia M 2019 Application of CSA-VMD and optimal scale morphological slice bispectrum in enhancing outer race fault detection of rolling element bearings *Mech. Syst. Signal Process.* **122** 56–86

[25] Guo J, Zhang H, Zhen D, Shi Z, Gu F and Ball A D 2020 An enhanced modulation signal bispectrum analysis for bearing fault detection based on non-Gaussian noise suppression *Measurement* **151** 107240

[26] Anjaneyulu L, Murthy N S and Sarma N V S N 2009 A novel method for recognition of modulation code of LPI radar signals *Int. J. Recent Trends Eng.* **1** 176

[27] Buades A, Coll B and Morel J-M 2005 A review of image denoising algorithms, with a new one *Multiscale Model. Simul.* **4** 490–530

[28] Sacchi M D, Ulrych T J and Walker C J 1998 Interpolation and extrapolation using a high-resolution discrete Fourier transform *IEEE Trans. Signal Process.* **46** 31–8

[29] Mendel J M 1991 Tutorial on higher-order statistics (spectra) in signal processing and system theory: theoretical results and some applications *Proc. IEEE* **79** 278–305

[30] Kachenoura A, Albera L, Bellanger J-J and Senhadji L 2008 Nonminimum phase identification based on higher order spectrum slices *IEEE Trans. Signal Process.* **56** 1821–9

[31] Sanaullah M 2013 A review of higher order statistics and spectra in communication systems *J. Sci. Front. Res. Phys. Space Sci.* **13** 31–50

[32] Pmod AD1. Digilent https://digilentinc.com/Products/Detail.cfm?NavPath=2,401,499%5C&Prod=PMOD-AD1

[33] Domínguez-Rodríguez I *et al* 2010 Sistema de digitalización de señal basado en FPGA y configurado utilizando MatLab *Científica* **14** 129–35

[34] Spartan-3E Starter Kit. Xilinx http://www.xilinx.com/products/boards-and-kits/HW-SPAR3E-SK-US-G.htm

[35] Timmer J, Lauk M and Deuschl G 1996 Quantitative analysis of tremor time series *Electroencephalogr. Clin. Neurophysiol. Electromyogr. Motor Control* **101** 461–8

[36] Palomino E 2007 *Elementos de Medición y Análisis de Vibraciones en Máquinas Rotatorias* Universidad Technolólogia de la Habana José Antonio Echeverría

[37] Ernesto H M F 2006 Diagnóstico del estado de cojinetes de rodamientos utilizando procesamiento cicloestacionario basado en cumulantes *Doctoral Thesis* Instituto Superior Politécnico José Antonio Echeverría, CUJAE, La Habana

[38] Akaike H 1974 A new look at the statistical model identification *IEEE Trans. Autom. Control* **19** 716–23

[39] Sequeira S 2011 Energy based spectrum sensing for enabling dynamic spectrum access in cognitive radios *Master's Thesis* Rutgers, The State University of New Jersey

[40] Wang X and Poor H V 1998 Blind multiuser detection: a subspace approach *IEEE Trans. Inf. Theory* **44** 677–90

[41] Wax M and Kailath T 1985 Detection of signals by information theoretic criteria *IEEE Trans. Acoust. Speech Signal Process.* **33** 387–92

[42] Grimaldi G and Manto M 2010 Neurological tremor: sensors, signal processing and emerging applications *Sensors* **10** 1399–422

[43] Abreu F, Montero F E, Suarez J R and Medina A 2012 Sistema para el análisis cualitativo del temblor humano *Congreso Internacional COMPUMAT 2012 (Havana, Cuba)*

[44] Domínguez Rodríguez I, Suárez Rodríguez J R and Hernández Montero F E 2013 Sistema para la cuantificación del temblor humano *V Latin American Congress on Biomedical Engineering CLAIB 2011 (Havana Cuba, 16–21 May 2011)* ed J F Méndez, T Y Aznielle Rodríguez, C F Calderón Marín, S B Llanusa Ruiz, J C Medina, H V Vázquez, M C Barreda and R R Rojas (Berlin: Springer) pp 658–61

[45] Sprdlik O 2012 Detection and estimation of human movement using inertial sensors: applications in neurology *PhD Thesis* Czech Technical University in Prague, Faculty of Electrical Engineering Department of Control Engineering https://support.dce.felk.cvut.cz/mediawiki/images/8/81/Diz_2012_sprdlik_otakar.pdf

[46] Graham B B 2000 Using an accelerometer sensor to measure human hand motion *Master's Thesis* Massachusetts Institute of Technology

[47] Engin M 2007 A recording and analysis system for human tremor *Measurement* **40** 288–93

[48] Fendy S 2008 Welch based denoising technique for a set of chirp signals corrupted by Gaussian noises *ITB J* **2** 115–29

Appendix A

Properties of cumulants

Cumulants can be used as an operator, in the same way that the statistical expectation operator $E\{\}$ is treated. The main properties of cumulants, which support this statement, are the following [1–3]:

1. Cumulants of signals or scaled processes, these scale factors not being random, correspond to the product of all these factors by the cumulants of the unscaled process:

$$c_n^x\{\lambda_0 \cdot x(k), \dots, \lambda_{N-1} \cdot x(k-k_{N-1})\} = \left(\prod_{i=0}^{N-1} \lambda_i\right) \cdot c_n^x\{x(k), \dots, x(k-k_{N-1})\}, \quad \text{(A.1)}$$

where λ_i y $c_n^x\{x(k), \dots, x(k-k_{N-1})\}$.

2. The cumulants are symmetric with respect to the position of their arguments $k_0 = 0$,

$$c_n^x\{x(k-k_0), \dots, x(k-k_{N-1})\} = c_n^x\{x(k-k_{i_0}), \dots, x(k-k_{i_{n-1}})\}, \quad \text{(A.2)}$$

where $(i_0, \dots, i_{N-1})$ is a permutation of $0, 1, \dots, N-1$. This means that the arguments of the cumulants can be exchanged without changing their value. In this way the fourth-order cumulants verify

$$c_4^x(\tau_1, \tau_2, \tau_3) = c_4^x(\tau_3, \tau_1, \tau_2) = c_4^x(\tau_2, \tau_3, \tau_1). \quad \text{(A.3)}$$

3. The cumulants are additive with respect to their arguments; that is, the cumulants of a sum of statistically independent arguments correspond to the sum of the cumulants of each individual argument,

$$\begin{aligned} &c_n^{x+y}\{x(k)+y(k), x(k-k_1), \dots, x(k-k_{N-1})\} \\ &= c_n^x\{x(k), x(k-k_1), \dots, x(k-k_{N-1})\} \\ &+ c_n^y y(k), x(k-k_1), \dots, x(k-k_{N-1}). \end{aligned} \quad \text{(A.4)}$$

4. Cumulants are invariant with respect to the addition of constants. With δ being a constant, it is verified that

$$\begin{aligned} &c_n^x\{\delta + x(k), x(k - k_1), \dots, x(k - k_{N-1})\} \\ &\quad = c_n^x\{x(k), x(k - k_1), \dots, x(k - k_{N-1})\} \\ &+ c_n^x\{y(k), \dots, y(k - k_{N-1})\}. \end{aligned} \tag{A.5}$$

5. If two random processes $x(k)$ and $y(k)$ are statistically independent, the cumulants of the two processes add, taking the value of the sum of the cumulants of each process separately; that is,

$$\begin{aligned} c_n^{x+y}\{x(k) + y(k), \dots, x(k - k_{N-1}) \cdot y(k - k_{N-1})\} &= c_n^x\{x(k), \dots, x(k - k_{N-1})\} \\ &+ c_n^y y(k), \dots, x(k - k_{N-1}). \end{aligned} \tag{A.6}$$

   Note that if the processes $x(k)$, $y(k)$ were not independent, according to property (3), $2n$ terms would appear on the right-hand side of this last expression.

6. If a subset of r arguments ($r \leqslant n$) are independent of the rest, then it holds that

$$c_n^x(x_1, \dots, x_n) = 0. \tag{A.7}$$

7. Cumulants of order n present $n!$ regions of symmetry,

$$\begin{aligned} &c_n(k_1, k_2, \dots, k_{N-1}) = c_n(k_2, k_1, \dots, k_{N-1}) \\ &\qquad = \cdots = c_n(k_{N-1}, k_{N-2}, \dots, k_1) = \\ &c_n(-k_1, k_2 - k_1, \dots, k_{N-1} - k_1) = \cdots \\ &\qquad = c_n(k_{N-1} - k_1, k_{N-2} - k_1, \dots, -k_1) = \\ &c_n(k_1 - k_{N-1}, k_2 - k_{N-1}, \dots, -k_{N-1}) = \cdots \\ &\qquad = c_n(-k_{N-1}, k_{N-2}, -k_1, \dots, k_1 - k_{N-1}). \end{aligned} \tag{A.8}$$

   But, contrary to what happens in the domain of the autocorrelation function, even symmetry does not generally hold for $n \geqslant 3$; that is,

$$c_n(k_1, k_2, \dots, k_{N-1}) \neq c_n(-k_1, -k_2, \dots, -k_{N-1}). \tag{A.9}$$

   Note that for the second-order case it reduces to $c_2(k_1) = c_2(-k_1)$ and even symmetry is satisfied. Let $v(k)$ be a Gaussian process independent of $x(k)$ (white or colored), which has degraded the process $x(k)$ originating the process $y(k) = x(k) + v(k)$; then for $n \geqslant 3$ it is verified that

$$c_n^y(k_1, k_2, \dots, k_{N-1}) = c_n^x(k_1, k_2, \dots, k_{N-1}), \tag{A.10}$$

while in the classical second-order case one has

$$c_n^y(k_1) = c_n^x(k_1) + c^v(k_1). \tag{A.11}$$

This last feature shows the greater robustness of higher-order statistics compared with the classical autocorrelation function, even when this noise is colored. Consequently, cumulants can obtain information from non-Gaussian processes without being affected by the presence of Gaussian noise and, therefore, obtain much more effective signal-to-noise ratios.

References

[1] Swami A and Mendel J M 1988 Cumulant-based approach to the harmonic retrieval problem *ICASSP-88, Int. Conf. on Acoustics, Speech, and Signal Processing* **4** 2264–7

[2] Mendel J M 1991 Tutorial on higher-order statistics (spectra) in signal processing and system theory: theoretical results and some applications *Proc. IEEE* **79** 278–305

[3] Swami A, Giannakis G B and Zhou G 1997 Bibliography on higher-order statistics *Signal Process.* **60** 65–126

IOP Publishing

Algorithms for Noise Reduction in Signals
Theory and practical examples based on statistical and convolutional analysis
Miguel Enrique Iglesias Martínez, Miguel Ángel García March, Carles Milián Enrique and Pedro Fernández de Córdoba

Appendix B

Moments, cumulants, and higher-order spectra

B.1 Moments

Let x be a given random variable. The moment of order k of the random variable x is defined as

$$\begin{aligned} m_k^x &= E\lfloor x^k \rfloor \\ &= (-j)^k \frac{\partial \Phi_x(\omega)}{\partial \omega_k}\bigg|_{\omega=0}, \end{aligned} \tag{B.1}$$

where $\Phi_x(\omega)$ is the characteristic function of the random variable x defined as

$$\Phi_x(\omega) = E[\exp(j\omega x)]. \tag{B.2}$$

From equations (B.1) and (B.2), the first-order moment of x is $m_1^x = E[x]$, the second-order moment of the random variable x is $m_2^x = E[x]$, and so on. The moment of order $k(k = k_1 + k_2)$ of two random variables x_1 and x_2 is defined as

$$E[x_1^{k1} x_2^{k2}] = (-j)^{k_1+k_2} \frac{\partial^{k_1} \partial^{k_2} \Phi_{x_1 x_2}(\omega_1, \omega_2)}{\partial \omega_1^{k_1} \partial \omega_2^{k_2}}\bigg|_{\omega_1=\omega_2=0}. \tag{B.3}$$

In general, the moment of order k of N random variables is defined as

$$\begin{aligned} m_k^N &= E[x_1^{k_1} x_2^{k_2}, \ldots, x_N^{k_N}] \\ &= (-j)^k \frac{\partial^k \Phi(\omega_1, \omega_2, \ldots, \omega_N)}{\partial \omega_1^{k_1} \partial \omega_2^{k_2}, \ldots, \partial \omega_N^{k_N}}\bigg|_{\omega_1=\omega_2=\cdots=\omega_N=0}, \end{aligned} \tag{B.4}$$

where $k = k_1 + k_2 + \cdots + k_N$ and the characteristic function is

$$\Phi(\omega_1, \omega_2, \ldots, \omega_N) = E[\exp(j\omega_1 x_1 + j\omega_2 x_2 + \cdots + j\omega_N x_N)]. \tag{B.5}$$

The higher-order moments ($k \geqslant 2$) are useful for characterizing random processes that are discrete in time. Then the nth-order moment of a random process $x(m)$ is defined as

$$m_k^x(\tau_1, \tau_2, \dots, \tau_{k-1}) = E[x(m) \cdot x(m+\tau_1) \cdot x(m+\tau_2) \cdot \cdots x(m+\tau_{k-1})]. \quad \text{(B.6)}$$

B.2 Cumulants

Cumulants are functions similar to moments. The difference is that the moments of a random process are obtained from the characteristic function $\Phi_x(\omega)$ and the generating function of cumulants is defined as the logarithm of the function feature $\Phi_x(\omega)$:

$$c_x(\omega) = \ln(\Phi_x(\omega)) = \ln(E[\exp(j\omega x)]). \quad \text{(B.7)}$$

Expanding in Taylor series the term $E[\exp(j\omega x)]$ in equation (B.7), the generating function of cumulants can be expressed as

$$c_x(\omega) = \ln\left(1 + m_1^x(j\omega) + \frac{m_2^x}{2!}(j\omega)^2 + \frac{m_3^x}{3!}(j\omega)^3 + \cdots + \frac{m_k^x}{k!}(j\omega)^k + \cdots\right), \quad \text{(B.8)}$$

where $m_k^x = E\lfloor x^k \rfloor$ is the moment of order of the random variable x. The cumulant of order k of the random variable is defined as

$$c_k^x = (-j)^k \frac{\partial^k c_x(\omega)}{\partial \omega_k}\Bigg|_{\omega=0}. \quad \text{(B.9)}$$

From equations (B.8) and (B.9) we have

$$\begin{aligned} c_1^x &= m_1^x \\ c_2^x &= m_2^x - (m_1^x)^2 \\ c_3^x &= m_3^x - 3m_1^x m_2^x + 2(m_1^x)^2, \end{aligned} \quad \text{(B.10)}$$

and, in general, the cumulant of order k of N random processes is defined as

$$c_k^N = (-j)^{k_1+\cdots+k_N} \frac{\partial^{k_1+\cdots+k_N} \ln \Phi_x(\omega_1, \omega_2, \dots, \omega_N)}{\partial \omega_1^{k_1} \partial \omega_2^{k_2} \cdots \partial \omega_N^{k_N}}\Bigg|_{\omega_1=\omega_2=\cdots=\omega_N=0}. \quad \text{(B.11)}$$

The cumulants of a random process $x(m)$ of zero mean value are given by

$$c_1^x = E[x(k)] = m_1^x = 0, \quad \text{(B.12)}$$

$$\begin{aligned} c_2^x &= E[x(m) \cdot x(m+k)] - E[x(m)]^2 \\ &= m_2^x - (m_1^x)^2 = m_x^2, \end{aligned} \quad \text{(B.13)}$$

$$\begin{aligned} c_3^x(k_1, k_2) &= m_3^x(k_1, k_2) - m_1^x[m_2^x(k_1) + m_2^x(k_2) + m_2^x(k_1 - k_2)] + 2(m_1^x)^3 \\ &= m_3^x(k_1, k_2), \end{aligned} \quad \text{(B.14)}$$

$$\begin{aligned} c_4^x(k_1, k_2, k_3) = {} & m_4^x(k_1, k_2, k_3) \\ & - m_2^x(k_1) \cdot m_2^x(k_3 - k_2) - m_2^x(k_2) \cdot m_2^x(k_3 - k_1) \\ & - m_2^x(k_3) \cdot m_2^x(k_2 - k_1) \\ & - m_1^x[m_3^x(k_2 - k_1, k_3 - k_1) + m_3^x(k_2, k_3) + m_3^x(k_2, k_4) \\ & \quad + m_3^x(k_1, k_2)] \\ & + (m_2^x)^2[m_1^x(k_1) + m_2^x(k_2) + m_2^x(k_3) + m_2^x(k_3 - k_1) \\ & \quad + m_2^x(k_3 - \tau_2) \\ & + m_2^x(k_2 - k_1)] - 6(m_1^x)^4, \end{aligned} \tag{B.15}$$

$$\begin{aligned} c_4^x(k_1, k_2, k_3) = {} & m_4^x(k_1, k_2, k_3) \\ & - m_2^x(k_1) \cdot m_2^x(k_3 - k_2) - m_2^x(k_2) \cdot m_2^x(k_3 - k_1) \\ & - m_2^x(k_3) \cdot m_2^x(k_2 - k_1). \end{aligned} \tag{B.16}$$

The cumulant of order n of a stationary random process, $x(m)$, can be written as

$$c_n^x(k_1, k_2, \dots, k_{n-1}) = m_n^x(k_1, k_2, \dots, k_{n-1}) - m_n^G(k_1, k_2, \dots, k_{n-1}) \tag{B.17}$$

for $n = 3, 4, \dots$ where $m_n^G(k_1, k_2, \dots, k_{n-1})$ is the moment of order n of a process with Gaussian distribution, which has the same value mean and autocorrelation function as the random process $x(m)$. From equation (B.17), it is evident that for a process with Gaussian distribution the cumulants for orders greater than 2 are equal to zero, since $m_n^x(k_1, k_2, \dots, k_{n-1}) = m_n^G(k_1, k_2, \dots, k_{n-1})$, so $c_n^x(k_1, k_2, \dots, k_{n-1}) = 0$.

B.3 Higher-order spectra

The spectrum of order k for a discrete random vector $X(k)$ is defined as the Fourier transform of dimension $k - 1$ of the sequence of cumulants of order k which results in the following way:

$$C_k^x(\omega_1, \omega_2, \dots, \omega_{n-1}) = \frac{1}{(2\pi)^{k-1}} \sum_{\tau_1=-\infty}^{\infty} \sum_{\tau_{n-1}=-\infty}^{\infty} c_k^x(\tau_1, \tau_2, \dots, \tau_{k-1}) e^{-j(w_1\tau_1 + \dots + w_{k-1}\tau_{k-1})}. \tag{B.18}$$

For the cases of the power spectrum ($k = 2$), it is defined as

$$C_2^x(w) = \frac{1}{2\pi} \sum_{\tau=-\infty}^{\infty} c_2^x(\tau) e^{-jw\tau}. \tag{B.19}$$

The bispectrum ($k = 3$) is

$$C_3^x(w_1, w_2) = \frac{1}{(2\pi)^2} \sum_{\tau_1=-\infty}^{\infty} \sum_{\tau_2=-\infty}^{\infty} c_3^x(\tau_1, \tau_2) e^{-j(w_1\tau_1 + w_2\tau_2)}, \tag{B.20}$$

and the trispectrum ($k = 4$) is

$$C_4^x(w_1, w_2, w_3) = \frac{1}{(2\pi)^3} \sum_{\tau_1=-\infty}^{\infty} \sum_{\tau_2=-\infty}^{\infty} \sum_{\tau_3=-\infty}^{\infty} c_4^x(\tau_1, \tau_2, \tau_3) e^{-j(w_1\tau_1 + w_2\tau_2 + w_3\tau_3)}. \tag{B.21}$$

Miguel Enrique Iglesias Martínez, Miguel Ángel García March, Carles Milián Enrique and Pedro Fernández de Córdoba

Appendix C

Calculation of the one-dimensional component of the fourth-order cumulative of a harmonic signal

Let $x(t) = \sum_{k=1}^{N} A_k \cos(w_k t + \phi_k)$ be a stationary real signal with zero mean value made up of a sum of N harmonics, of which $w_c = 2\pi f_c$ is the frequency of the signal; ϕ_k is the phase, a statistically independent and uniformly distributed random variable of $0 \leqslant \phi \leqslant 2\pi$ with pdf $= \frac{1}{2\pi}$; and A_k is the amplitude of each harmonic. Then, the fourth-order cumulant of $x(t)$ can be expressed as follows:

$$\begin{aligned} c_4^x(\tau_1, \tau_2, \tau_3) = {} & m_4^x(\tau_1, \tau_2, \tau_3) \\ & - m_2^x(\tau_1) \cdot m_2^x(\tau_3 - \tau_2) - m_2^x(\tau_2) \cdot m_2^x(\tau_3 - \tau_1) \\ & - m_2^x(\tau_3) \cdot m_2^x(\tau_2 - \tau_1). \end{aligned} \tag{C.1}$$

Then if only the one-dimensional component of the fourth-order cumulant is used, equation (C.1) is as follows:

$$c_4^x(\tau_1, 0, 0) = E\{x(t)^3 \cdot x(t + \tau_1)\} - 3 \cdot E\{x(t) \cdot x(t + \tau_1)\} \cdot E\{x^2(t)\}. \tag{C.2}$$

For $x(t)$, the second-order cumulant $E\{x(t) \cdot x(t + \tau_1)\}$ is equal to $\sum_{k=1}^{N} \frac{A_k^2}{2} \cos(w_k \tau)$ (see appendix D) and, for a stationary process with zero mean and the term $E\{x^2(t)\}$, corresponds to the variance of the given signal that for the treated harmonic signal is equivalent to $\sum_{k=1}^{N} \frac{A_k^2}{2}$. Then expanding the left term of the subtraction by applying the trigonometric identity of the cubic cosine product ($\cos^3 \theta = \frac{1}{4}(3\cos(\theta) + \cos(3\theta))$) and taking $w_k t + \phi_k = \theta$, we have the following:

$$
\begin{aligned}
E\{x(t)^3 \cdot x(t+\tau_1)\} &= E\left\{\sum_{k=1}^{N} A_k \cos(w_k t + \phi_k)^3 \cdot \sum_{k=1}^{N} A_k \cos(w_k t + w_k \tau_1 + \phi_k)\right\} \\
&= \sum_{k=1}^{N} A_k^4 E\{\cos^3(\theta) \cdot \cos(\theta + w_k\tau_1)\} \\
&= \sum_{k=1}^{N} A_k^4 E\left\{\left[\frac{3}{4}\cos(\theta) + \frac{1}{4}\cos(3\theta)\right] \cdot \cos(\theta + w_k\tau_1)\right\} \\
&= \sum_{k=1}^{N} A_k^4 E\left\{\frac{3}{4}\cos(\theta) \cdot \cos(\theta + w_k\tau_1) + \frac{1}{4}\cos(3\theta)\right. \\
&\qquad\qquad \left. \cdot \cos(\theta + w_k\tau_1)\right\} \\
&= \sum_{k=1}^{N} \frac{A_k^4}{4} E\{3\cos(\theta)\cos(\theta + w_k\tau_1) + \cos(3\theta)\cos(\theta + w_k\tau_1)\} \\
&= \sum_{k=1}^{N} \frac{A_k^4}{8} E\{3[\cos(-w_k\tau_1) + \cos(2\theta + w_k\tau_1)] \\
&\qquad\qquad + [\cos(2\theta - w_k\tau_1) + \cos(4\theta + w_k\tau_1)]\} \\
&= \sum_{k=1}^{N} \frac{3A_k^4}{8} \cos(w_k\tau_1).
\end{aligned} \tag{C.3}
$$

By auxiliary calculations applying the identity of the product of cosines and substituting in (C.3), we have

$$
\begin{aligned}
\cos(\theta) \cdot \cos(\theta + w_k\tau_1) &= \frac{1}{2}\cos(\theta - \theta - w_k\tau_1) + \frac{1}{2}\cos(\theta + \theta + w_k\tau_1) \\
&= \frac{1}{2}\cos(-w_k\tau_1) + \frac{1}{2}\cos(2\theta + w_k\tau_1)
\end{aligned} \tag{C.4}
$$

and

$$
\begin{aligned}
\cos(3\theta) \cdot \cos(\theta + w_k\tau_1) &= \frac{1}{2}\cos(3\theta - \theta - w_k\tau_1) + \frac{1}{2}\cos(3\theta + \theta + w_k\tau_1) \\
&= \frac{1}{2}\cos(2\theta - w_k\tau_1) + \frac{1}{2}\cos(4\theta + w_k\tau_1).
\end{aligned} \tag{C.5}
$$

Substituting the auxiliary results of (C.3) in equation (C.2), we have the following:

$$
\begin{aligned}
c_4^x(\tau_1, 0, 0) &= \sum_{k=1}^{N} \frac{3A_k^4}{8}\cos(w_k\tau_1) - 3\sum_{k=1}^{N}\frac{A_k^2}{2} \cdot \sum_{k=1}^{N}\frac{A_k^2}{2}\cos(w_k\tau) \\
&= \sum_{k=1}^{N} \frac{3A_k^4}{8}\cos(w_k\tau_1) - \sum_{k=1}^{N}\frac{3A_k^4}{4}\cos(w_k\tau) \\
&= \sum_{k=1}^{N} - \frac{3A_k^4}{8}\cos(w_k\tau_1),
\end{aligned} \tag{C.6}
$$

which matches reference [1] making $\tau_1 = \tau_2 = \tau_3 = 0$.

Reference

Mendel J M 1991 Tutorial on higher-order statistics (spectra) in signal processing and system theory: theoretical results and some applications *Proc. IEEE* **79** 278–305

Appendix D

Calculation of the autocorrelation function of a harmonic signal

Let $x(t) = \sum_{k=1}^{N} A_k \cos(w_k t + \phi_k)$ be a stationary real signal with zero mean value made up of a sum of N harmonics, of which $w_c = 2\pi f_c$ is the frequency of the signal; ϕ_k is the phase, a statistically independent and uniformly distributed random variable of $0 \leqslant \phi \leqslant 2\pi$ with pdf $= \frac{1}{2\pi}$; and A_k is the amplitude of each harmonic. Then the second-order cumulant of $x(t)$ can be expressed as follows:

$$E[x(t)] = x(t) \cdot x(t + \tau_1). \tag{D.1}$$

Then substituting in equation (D.1) and expanding the function, applying the trigonometric identity of the cosine product $\cos A \cdot \cos B = \frac{1}{2}\cos(A - B) + \frac{1}{2}\cos(A + B)$, we have the following:

$$\begin{aligned}
x(t) \cdot x(t + \tau_1) &= E\left\{\sum_{k=1}^{N} A_k \cos(w_k t + \phi_k) \sum_{k=1}^{N} A_k \cos(w_k t + w_k \tau_1 + \phi_k)\right\} \\
&= E\left\{\sum_{k=1}^{N} A_k^2 \cos(w_k t + \phi) \cdot \cos(w_k t + w_k \tau_1 + \phi_k)\right\} \\
&= E\left\{\sum_{k=1}^{N} \frac{A_k^2}{2} \cos(w_k t + \phi) \cdot \cos(w_k t + w_k \tau_1 + \phi_k)\right\} \\
&= E\left\{\sum_{k=1}^{N} \frac{A_k^2}{2} [\cos(-w_k \tau_1) + \cos(2\theta + w_k \tau_1)]\right\} \\
&= E\left\{\sum_{k=1}^{N} \frac{A_k^2}{2} \cos(-w_k \tau_1)\right\} \\
&= \frac{1}{2\pi} \int_0^{2\pi} \sum_{k=1}^{N} \frac{A_k^2}{2} \cos(-w_k \tau_1) d\phi_k,
\end{aligned}$$

doi:10.1088/978-0-7503-3591-1ch7

and

$$E\{x(t) \cdot x(t + \tau_1)\} = \sum_{k=1}^{N} \frac{A_k^2}{2} \cos(w_k \tau_1). \tag{D.2}$$

By auxiliary calculations applying the identity of the product of cosines, we have

$$\begin{aligned} \cos(\theta) \cdot \cos(\theta + w_k\tau_1) &= \frac{1}{2}\cos(\theta - \theta - w_k\tau_1) \\ &\quad + \frac{1}{2}\cos(\theta + \theta + w_k\tau_1) \\ &= \frac{1}{2}\cos(-w_k\tau_1) + \frac{1}{2}\cos(2\theta + w_k\tau_1). \end{aligned} \tag{D.3}$$

Equation (D.3) corresponds to the identity of the product of cosines.

IOP Publishing

Algorithms for Noise Reduction in Signals
Theory and practical examples based on statistical and convolutional analysis
Miguel Enrique Iglesias Martínez, Miguel Ángel García March, Carles Milián Enrique and Pedro Fernández de Córdoba

Appendix E

Examples of codes

E.1: Matlab convolution function

```
function [Cum_conv]=func_conv(u,v)

m=length(u);
n=length(v);
%%%%%%%%%%%%%%%%%%%%% CONVOLUTION %%%%%%%%%%%%%%%%%%%%%

for k = 1:n                    %tau INDEX

F = 0;

     for j = max(1,k+1-n): min(k,m)

          F = F + u(j)*v(k+1-j);

     end

     Cum_conv(k) = F/min(k,m);
end
```

E.2: Matlab function for the one-dimensional component of fourth order cumulant

```
function [C4x_uv]=func_cum4uni_vertical(x)

x = x - mean(x) ;

N=length(x)

%%%%%%%%ONE-DIMENSIONAL COMPONENT%%%%%%%%%
%%%%%%%%%% OF THE FOURTH ORDER CUMULANT%%%%%
%%%%%%%%%%%E{x(k)^3*x(k+m)}%%%%%%%%%%%%%%%%
```

doi:10.1088/978-0-7503-3591-1ch8

```
for m = 1:N                        %tau  INDEX

F = 0;                             %E{}
F1=0;                              %Rxx

    for k = 1:(N+1-m)

        F = F + x(k)*x(k)*x(k)*x(k-1+m);
      F1 = F1 + x(k)*x(k-1+m);
    end

    C4xx(m) = F/(N+1-m);
   Rxx(m) = F1/(N+1-m);
end

C4x_uv=C4xx-3*Rxx*Rxx(1)   %OUTPUT
```

E.3: Example using a harmonic signal

```
%%%%%%%%%%%%%%%%%%%%%%%%%%%%%%%%%%%%%%%%%%%%%
%%%%%%%%%%%%%% Harmonic Signal %%%%%%%%%%%%%%

clear all;
close all;
xt=0;
px=0;
pn=0;
pxot=0;
pxit=0;
Fs =1000;                          % Sampling frequency
fc=50;
T = 1/Fs;                          % Sample time
KB=1024
L = 1*KB                           % Length of signal
t = (0:L-1)*T;                     % Time vector
A=1;                               % amplitude
nfft=2^nextpow2(L);
wc=2*pi*fc;


xt=cos(2*pi*fc*t+pi/4);            %cosine signal


figure(1);
plot(t,xt,'k');
title('Useful Signal')
```

```
xlabel('Time(s)')
ylabel('Amplitude(u)')
grid on;

xt_fft=fft(xt,nfft)/L;
f=Fs*linspace(0,1,nfft);

figure(2)
plot(f,abs(xt_fft),'k')
xlabel('Frequency(Hz)')
ylabel('Amplitude(u)')
title('Useful Signal Spectrum')
grid on;



stream=RandStream('mrg32k3a');
noise=2.2*randn(stream,1,length(xt))';      % contains exactly the same
    values as noisem
xt_noise=xt+noise';


figure(3);
plot(t,xt_noise,'k');
title('Noisy Signal')
xlabel('Time(s)')
ylabel('Amplitude(u)')
grid on;

xt_noise_fft=fft(xt_noise,nfft)/L;    %Fourier Transform of the Noisy
    Signal

figure(4)
plot(f,abs(xt_noise_fft),'k')
xlabel('Frequency(Hz)')
ylabel('Amplitude(u)')
title('Noisy Signal Spectrum')
grid on;



%%%%%%%%%%%%%%%%%%%%%%%%%%%%%%%%%%%%%%%%%%%%%%%%%%%%%
%%%%%%%%%%%%%%%%%%% C4-CONV-IFFT %%%%%%%%%%%%%%%%%%%
%%%%%%%%%%%%%%%%%%%%%%%%%%%%%%%%%%%%%%%%%%%%%%%%%%%%%
tic
C4x_uv=func_cum4uni_vertical(xt_noise);  % one-dimensional component of
```

```
    the fourth order cumulant

Cum_conv=func_conv(C4x_uv,xt_noise);
Cum_conv=(-1)*Cum_conv;

conv_fft=fft(Cum_conv,nfft)/L;

ajuste=((16/3)*abs(conv_fft)).^(1/5);     %amplitude adjustment

ajuste_mean=mean(ajuste);

for qq=1:1                                %correction by mean
 for q=1:length(ajuste)
  ajuste(q)=ajuste(q)-ajuste_mean;
  if ajuste(q)<0
      ajuste(q)=0;
  end;
 end
end



nuevo_num = ajuste.*exp(i*angle(conv_fft));   %combining the new number

ajuste_ifft=ifft(nuevo_num,nfft)*L;           %antitransformation

atn=real(ajuste_ifft);

mod_atn=2*atn.*atn;                       %envelope detection
b=firpm(20, [0 0.03 0.1 1], [1 1 0 0]);
sal=filter(b,1,mod_atn);
sal=sqrt(sal);
atn_final=atn./sal;

t_cum=toc

figure(5);
plot(t,xt,'b');
hold on
plot(t,atn_final,'r');
title('Comparison in time Useful Signal- Proposed algorithm')
xlabel('Time(s)')
ylabel('Amplitude(u)')
grid on;
legend('Useful','Output',2)
```

```
atn_final_fft=fft(atn_final,nfft)/L;

figure(6)
plot(f,abs(xt_fft),'b')
hold on
plot(f,abs(atn_final_fft),'r')
xlabel('Frequency(Hz)')
ylabel('Amplitude(u)')
title('Comparison of the spectrum of the Useful Signal-Proposed
    algorithm')
grid on;
legend('Useful','Output')

figure(7)
plot(f,abs(xt_noise_fft),'b')
hold on
plot(f,abs(atn_final_fft),'r')
xlabel('Frequency(Hz)')
ylabel('Amplitude(u)')
title('Comparison of the spectrum of the Noisy Signal-Proposed
    algorithm')
grid on;
legend('Noisy','Output',2)





for i=1:length(xt)
  px=px+((abs(xt(i)))^2/length(xt));
end

for i=1:length(noise)
  pn=pn+((abs(noise(i)))^2/length(noise));
end

SNR=10*log(px/pn)


for i=1:length(atn_final)
  pxot=pxot+((abs(atn_final(i)))^2/length(atn_final)); ;
end

pno=pxot-px;
SNRO=10*log(px/pno)

CORRE_INICIAL=corrcoef(xt,xt_noise)
CORRE_FINAL=corrcoef(xt,atn_final)
```

E.4: Example using a rectangular pulse

```
%%%%%%%%%%%%%%%%%%%%%%%%%%%%%%%%%%%%%%%%%%%%%%%
%%%%%%%%%%%%% Rectangular Pulse %%%%%%%%%%%
%%%%%%%%%%%%%%%%%%%%%%%%%%%%%%%%%%%%%%%%%%%%%%%
clear all;
close all;
xt=0;
px=0;
pn=0;
pxot=0;
pxit=0;
Fs =10000;                    % Sampling frequency
fc=100;                       % fundamental frequency
T = 1/Fs;                     % Sample time
L = 8192                      % Length of signal
t = (0:L-1)*T;                % Time vector
A=1;                          % amplitude
nfft=2^nextpow2(L);
sxin=cos(2*pi*fc*t)';
wc=2*pi*fc;
for n=1:6
  xt=xt+(4*A/3)*sin((2*n-1)*wc*t+pi)/(2*n-1);
end

figure(1);
plot(t,xt,'k');
title('Useful Signal')
xlabel('Time(s)')
ylabel('Amplitude(u)')
grid on;

xt_fft=fft(xt,nfft)/L;
f=Fs*linspace(0,1,nfft);

figure(2)
plot(f,abs(xt_fft),'k')
xlabel('Frequency(Hz)')
ylabel('Amplitude(u)')
title('Useful Signal Spectrum')
grid on;


```

```

stream=RandStream('mrg32k3a');
noise=1.4*randn(stream,1,length(xt))';      % contains exactly the same values as noisem
xt_noise=xt+noise';


figure(3);
plot(t,xt_noise,'k');
title('Noisy Signal')
xlabel('Time(s)')
ylabel('Amplitude(u)')
grid on;

xt_noise_fft=fft(xt_noise,nfft)/L;         % calculation of the noisy signal spectrum

figure(4)
plot(f,abs(xt_noise_fft),'k')
xlabel('Frequency(Hz)')
ylabel('Amplitude(u)')
title('Noisy Signal Spectrum')
grid on;




%%%%%%%%%%%%%%%%%%%%%%%%%%%%%%%%%%%%%%%%%%%%%%%%%%%%%%%%%%%%%%%%%%%%%%%%%%%%%%%%%%%%%%%%%%%%%%%%%%%%
%%%%%%%%%%%%%%%%%%%%%%%%%%%%%%%%%%%% C4-CONV-IFFT %%%%%%%%%%%%%%%%%%%%%%%%%%%%%%%%%%%%
%%%%%%%%%%%%%%%%%%%%%%%%%%%%%%%%%%%%%%%%%%%%%%%%%%%%%%%%%%%%%%%%%%%%%%%%%%%%%%%%%%%%%%%%%%%%%%%%%%%%
tic
C4x_uv=func_cum4uni_vertical(xt_noise);  % one-dimensional component of the fourth order cumulant

Cum_conv=func_conv(C4x_uv,xt_noise);
Cum_conv=(-1)*Cum_conv;

conv_fft=fft(Cum_conv,nfft)/L;

ajuste=((16/3)*abs(conv_fft)).^(1/5);     %amplitude adjustment

ajuste_mean=mean(ajuste);

for qq=1:1                                 %correction by mean
 for q=1:length(ajuste)
  ajuste(q)=ajuste(q)-ajuste_mean;
```

```
  if ajuste(q)<0
      ajuste(q)=0;
  end;
 end
end



nuevo_num = ajuste.*exp(i*angle(conv_fft));   %combining the new number

ajuste_ifft=ifft(nuevo_num,nfft)*L;           %antitransformation

atn=real(ajuste_ifft);

mod_atn=2*atn.*atn;                           %envelope detection
b=firpm(20, [0 0.03 0.1 1], [1 1 0 0]);
sal=filter(b,1,mod_atn);
sal=sqrt(sal);
atn_final=atn./sal;

t_cum=toc

figure(5);
plot(t,xt,'b');
hold on
plot(t,atn_final,'r');
title('Comparison in time Useful Signal- Proposed algorithm')
xlabel('Time(s)')
ylabel('Amplitude(u)')
grid on;
legend('Useful','Output',2)

atn_final_fft=fft(atn_final,nfft)/L;

figure(6)
plot(f,abs(xt_fft),'b')
hold on
plot(f,abs(atn_final_fft),'r')
xlabel('Frequency(Hz)')
ylabel('Amplitude(u)')
title('Comparison of the spectrum of the Useful Signal-Proposed algorithm')
grid on;
legend('Useful','Output')
```

```
figure(7)
plot(f,abs(xt_noise_fft),'b')
hold on
plot(f,abs(atn_final_fft),'r')
xlabel('Frequency(Hz)')
ylabel('Amplitude(u)')
title('Comparison of the spectrum of the Noisy Signal-Proposed algorithm')
grid on;
legend('Noisy','Output',2)

for i=1:length(xt)
  px=px+((abs(xt(i)))^2/length(xt));
end

for i=1:length(noise)
  pn=pn+((abs(noise(i)))^2/length(noise));
end

SNR=10*log(px/pn)

for i=1:length(atn_final)
  pxot=pxot+((abs(atn_final(i)))^2/length(atn_final)); ;
end

pno=pxot-px;
SNRO=10*log(px/pno)

CORRE_INICIAL=corrcoef(xt,xt_noise)
CORRE_FINAL=corrcoef(xt,atn_final)
```

E.5: Example using a triangular pulse

```
%%%%%%%%%%%%%%%%%%%%%%%%%%%%%%%%%%%%%%%%%%%%%%%%%%%%%%%%%%%%%%%%%%%
%%%%%%%%%%%%%%%%%%%%%% Triangular Pulse %%%%%%%%%%%%%%%%%%%%%%%%%%%
%%%%%%%%%%%%%%%%%%%%%%%%%%%%%%%%%%%%%%%%%%%%%%%%%%%%%%%%%%%%%%%%%%%
clear all;
close all;
xt=0;
px=0;
pn=0;
pxot=0;
```

```
pxit=0;
Fs =5000;                      % Sampling frequency
fc=100;                        % Fundamental frequency
T = 1/Fs;                      % Sample time
L = 8192                       % Length of signal
t = (0:L-1)*T;                 % Time vector
A=1;                           % amplitude
nfft=2^nextpow2(L);
sxin=cos(2*pi*fc*t)';
wc=2*pi*fc;


xt=(8*A/(pi)^2)*(sin(wc*t)-sin(3*wc*t)/9 + sin(5*wc*t)/25-sin(7*wc*t)
    /49 + sin(9*wc*t)/81 - sin(11*wc*t)/121);


figure(1);
plot(t,xt,'b');
title('Useful Signal')
xlabel('Time(s)')
ylabel('Amplitude(u)')
grid on;

xt_fft=fft(xt,nfft)/L;
f=Fs*linspace(0,1,nfft);

figure(2)
plot(f,abs(xt_fft),'b')
xlabel('Frequency(Hz)')
ylabel('Amplitude(u)')
title('Useful Signal Spectrum')
grid on;



stream=RandStream('mrg32k3a');
noise=0.8*randn(stream,1,length(xt))';      % contains exactly the same
    values as noisem
xt_noise=xt+noise';


figure(3);
plot(t,xt_noise,'b');
title('Noisy Signal')
xlabel('Time(s)')
ylabel('Amplitude(u)')
```

```
grid on;

xt_noise_fft=fft(xt_noise,nfft)/L;          % calculation of the noisy signal spectrum

figure(4)
plot(f,abs(xt_noise_fft),'b')
xlabel('Frequency(Hz)')
ylabel('Amplitude(u)')
title('Noisy Signal Spectrum')
grid on;



%%%%%%%%%%%%%%%%%%%%%%%%%%%%%%%%%%%%%%%%%%%%%%%%%%%%%
%%%%%%%%%%%%%%%%%%% C4-CONV-IFFT %%%%%%%%%%%%%%%%%%%%
%%%%%%%%%%%%%%%%%%%%%%%%%%%%%%%%%%%%%%%%%%%%%%%%%%%%%
tic
C4x_uv=func_cum4uni_vertical(xt_noise);  % one-dimensional component of the fourth order cumulant

Cum_conv=func_conv(C4x_uv,xt_noise);
Cum_conv=(-1)*Cum_conv;

conv_fft=fft(Cum_conv,nfft)/L;

ajuste=((16/3)*abs(conv_fft)).^(1/5);     %amplitude adjustment

ajuste_mean=mean(ajuste);

for qq=1:1                                 %correction by mean
 for q=1:length(ajuste)
  ajuste(q)=ajuste(q)-ajuste_mean;
  if ajuste(q)<0
      ajuste(q)=0;
  end;
 end
end



nuevo_num = ajuste.*exp(i*angle(conv_fft));    %combinning the new number

ajuste_ifft=ifft(nuevo_num,nfft)*L;            %antitrasnformation

```

```
atn=real(ajuste_ifft);

mod_atn=2*atn.*atn;                                   %envelope detection
b=firpm(20, [0 0.03 0.1 1], [1 1 0 0]);
sal=filter(b,1,mod_atn);
sal=sqrt(sal);
atn_final=atn./sal;

t_cum=toc

figure(5);
plot(t,xt,'b');
hold on
plot(t,atn_final,'r');
title('Comparison in time Useful Signal- Proposed algorithm')
xlabel('Time(s)')
ylabel('Amplitude(u)')
grid on;
legend('Useful','Output',2)

atn_final_fft=fft(atn_final,nfft)/L;

figure(6)
plot(f,abs(xt_fft),'b')
hold on
plot(f,abs(atn_final_fft),'r')
xlabel('Frequency(Hz)')
ylabel('Amplitude(u)')
title('Comparison of the spectrum of the Useful Signal-Proposed
    algorithm')
grid on;
legend('Useful','Output')

figure(7)
plot(f,abs(xt_noise_fft),'b')
hold on
plot(f,abs(atn_final_fft),'r')
xlabel('Frequency(Hz)')
ylabel('Amplitude(u)')
title('Comparison of the spectrum of the Noisy Signal-Proposed
    algorithm')
grid on;
legend('Noisy','Output',2)
```

```
for i=1:length(xt)
  px=px+((abs(xt(i)))^2/length(xt));
end

for i=1:length(noise)
  pn=pn+((abs(noise(i)))^2/length(noise));
end

SNR=10*log(px/pn)


for i=1:length(atn_final)
  pxot=pxot+((abs(atn_final(i)))^2/length(atn_final)); ;
end

pno=pxot-px;
SNRO=10*log(px/pno)

CORRE_INICIAL=corrcoef(xt,xt_noise)
CORRE_FINAL=corrcoef(xt,atn_final)
```

E.6: Example using ADXL203 accelerometer

```
clear all;
close all;
px=0;
pn=0;
pxot=0;
pxit=0;
Fs=10000
sensor=load('movil.txt');          %SAMPLES SIGNAL

sxin=sensor;

sxin=sxin-mean(sxin);

stream=RandStream('mrg32k3a');
noise=0.42*randn(stream,1,length(sxin))';

sxnoise=sxin+noise;
L=length(sxin);
nfft=L;
T = 1/Fs;                          % Sample time
t = (0:L-1)*T;                     % Time vector
```

```
figure(1);
plot(t,sxin,'k');
title('Signal Sensor')
xlabel('Time(s)')
ylabel('Amplitude(u)')
grid on;

figure(2)
plot(t,sxnoise,'k');
title('Noisy Signal')
xlabel('Time(s)')
ylabel('Amplitude(u)')
grid on;


sxnoise_fft=fft(sxnoise,nfft)/length(sxnoise);
sxin_fft=fftshift(sxnoise_fft);
f=Fs*linspace(0,1,nfft);

figure(3)
plot(f,abs(sxnoise_fft),'k')
xlabel('Frequency(Hz)')
ylabel('Amplitude')
title('Noisy Signal Spectrum')
grid on;

figure(4)
w=[0:nfft-1]*pi/nfft;
plot(w/pi,abs(sxnoise_fft),'k');
xlabel('Frequency(rad/s)')
ylabel('Amplitude')
title('Useful Signal Spectrum in rad/s')
grid on;

figure(5)
fase=unwrap(angle(sxnoise_fft),'k')
plot(f,fase/pi);
title('Phase Spectrum')
grid on;


%%%%%%%%%%%%%%%%%%%%%%%%%%%%%%%%%%%%%%%%%%%%%%%%
%%%%%%%%%%%%%% C4-CONV-IFFT MODEL %%%%%%%%%%%%%%
%%%%%%%%%%%%%%%%%%%%%%%%%%%%%%%%%%%%%%%%%%%%%%%%
tic
```

```
C4x_uv=func_cum4uni_vertical(sxnoise);   %FOURTH ORDER CUMULANT

Cum_conv=func_conv(C4x_uv,sxnoise);      %PERFORMING THE CONVOLUTION
Cum_conv=(-1)*Cum_conv;

conv_fft=fft(Cum_conv,nfft)/L;

ajuste=((16/3)*abs(conv_fft)).^(1/5);    %AMPLITUDE CORRECTION

ajuste_mean=mean(ajuste);

for qq=1:2                               %CORRECTION BY MEAN
 for q=1:length(ajuste)
  ajuste(q)=ajuste(q)-ajuste_mean;
  if ajuste(q)<0
      ajuste(q)=0;
  end;
 end
end



nuevo_num = ajuste.*exp(i*angle(conv_fft));

ajuste_ifft=ifft(nuevo_num,nfft)*L;      %INVERSE TRANSFORMATION

atn=real(ajuste_ifft);


atn_final=atn;

t_cum=toc

figure(6);
plot(t,sxin,'b');
hold on
plot(t,atn_final,'r');
title(' Comparison between  the Useful Signal-Proposed Algorithm')
xlabel('Time(s)')
ylabel('Amplitude(u)')
grid on;
legend('Useful','Output')

atn_final_fft=fft(atn_final,nfft)/L;
sxin_fft=fft(sxin,nfft)/L;

```

```
figure(7)
plot(f,abs(sxin_fft),'b')
hold on
plot(f,abs(atn_final_fft),'r')
xlabel('Frequency(Hz)')
ylabel('Amplitude(u)')
title('Comparison between   the Useful Signal-Proposed Algorithm
    Spectrum')
grid on;
legend('Useful','Output')

figure(8)
plot(f,abs(sxnoise_fft),'b')
hold on
plot(f,abs(atn_final_fft),'r')
xlabel('Frequency(Hz)')
ylabel('Amplitude(u)')
title('Comparison between the Noisy Signal-Proposed Algorithm Spectrum')
grid on;
legend('Noisy','Output')



for i=1:length(sxin)
  px=px+((abs(sxin(i)))^2/length(sxin));
end

for i=1:length(noise)
  pn=pn+((abs(noise(i)))^2/length(noise));
end

SNR=10*log(px/pn)


for i=1:length(atn_final)
  pxot=pxot+((abs(atn_final(i)))^2/length(atn_final)); ;
end

pno=pxot-px;
SNRO=10*log(px/pno)

CORRE_INICIAL=corrcoef(sxin,sxnoise)
CORRE_FINAL=corrcoef(sxin,atn_final)
```

E.7: Example using a binary phase shift keying (BPSK) modulation

```
%%%%%%%%%%%%%%%%%%%%%% BPSK SIGNAL %%%%%%%%%%%%%%%%%%%%%%%%%
%%%%%%%%%%%%%%%%%%%%%%%%%%%%%%%%%%%%%%%%%%%%%%%%%%%%%%%%%%%%%%%%%%%%%%
clear all;
close all;
xt=0;
px=0;
pn=0;
pxot=0;
pxit=0;
%%%%%%%%%%%%%%%%%%%%%%%%%%%%%%%%%%%%%%%%%%%%%%%%%%%%%%%%
Fs = 20000;                          % Sampling frequency
fc=1000;                             % Fundamental frequency
moduladora=[0 0 1 0  1 0];           % sequence
[xt bit]=bpskd(moduladora,fc)
T = 1/Fs;                            % Sample time
L = length(xt);                      % Length of signal
t = (0:L-1)*T;                       % Time vector
nfft=length(bit);


figure(1);
plot(t,xt,'k');
title('Useful Signal')
xlabel('Time(s)')
ylabel('Amplitude(u)')
grid on;

xt_fft=fft(xt,nfft)/L;
f=Fs*linspace(0,1,nfft);

figure(2)
plot(f,abs(xt_fft),'k')
xlabel('Frequency(Hz)')
ylabel('Amplitude(u)')
title('Useful Signal Spectrum')
grid on;



%%%%%%%%%%%%%%%%%%%%%%%%%%%%%%%%%%%%%%%%%%%%%%%%%%%%%%
stream=RandStream('mrg32k3a');
noise=1.0*randn(stream,1,length(xt))';      % contains exactly the same
    values as noisem
xt_noise=xt+noise';


figure(3);
```

```
plot(t,xt_noise,'k');
title('Noisy Signal')
xlabel('Time(s)')
ylabel('Amplitude(u)')
grid on;

xt_noise_fft=fft(xt_noise,nfft)/L;

figure(4)
plot(f,abs(xt_noise_fft),'k')
xlabel('Frequency(Hz)')
ylabel('Amplitude(u)')
title('Noisy Signal Spectrum')
grid on;


reg_vent=0;
t_total = (0:length(xt_noise)-1)*T;          % Time vector
nfft_total=length(xt_noise);                 % FFT length total vector
L_total=length(xt_noise);
vent_sim=200;
num_iter=length(xt)/vent_sim;
sx_final=zeros(1,length(xt));
tic
for i=1:num_iter
   xt_noise_v=xt_noise(reg_vent+1:vent_sim+reg_vent);
   t = (0:length(xt_noise_v)-1)*T;          % Time vector
   nfft=length(xt_noise_v);                 % FFT length
   L=length(xt_noise_v);
   C4x_uv=func_cum4uni_vertical(xt_noise_v);
   Cum_conv=func_conv(C4x_uv,xt_noise_v);
   Cum_conv=(-1)*Cum_conv;


   sx_final(reg_vent+1:vent_sim+reg_vent)=Cum_conv;

   reg_vent=reg_vent+vent_sim;              %update the vector
end

%sx_final=(-1)*sx_final;
mod_atn=2*sx_final.*sx_final;
b=firpm(10, [0 0.03 0.1 1], [1 1 0 0]);
sal=filter(b,1,mod_atn);
sal=sqrt(sal);
atn_final=sx_final./sal;

t_cum=toc


figure(5);
```

```
plot(t_total,xt,'b');
hold on
plot(t_total,atn_final,'r');
title('Comparison in time Useful Signal-Proposed Algorithm')
xlabel('Time(s)')
ylabel('Amplitude(u)')
grid on;
legend('Useful','Output',2)

atn_final_fft=fft(atn_final,nfft_total)/L_total;

figure(6)
plot(f,abs(xt_fft),'b')
hold on
plot(f,abs(atn_final_fft),'r')
xlabel('Frequency(Hz)')
ylabel('Amplitude(u)')
title('Comparison of the spectrum Useful Signal-Proposed Algorithm')
grid on;
legend('Useful','Output')

figure(7)
plot(f,abs(xt_noise_fft),'b')
hold on
plot(f,abs(atn_final_fft),'r')
xlabel('Frequency(Hz)')
ylabel('Amplitude(u)')
title('Comparison of the spectrum Noisy Signal-Proposed Algorithm')
grid on;
legend('Noisy','Output',2)





for i=1:length(xt)
  px=px+((abs(xt(i)))^2/length(xt));
end

for i=1:length(noise)
  pn=pn+((abs(noise(i)))^2/length(noise));
end

SNR=10*log(px/pn)


for i=1:length(atn_final)
  pxot=pxot+((abs(atn_final(i)))^2/length(atn_final)); ;
end

pno=pxot-px;
```

```
SNRO=10*log(px/pno)

CORRE_INICIAL=corrcoef(xt,xt_noise)
CORRE_FINAL=corrcoef(xt,atn_final)
```

E.8: Example using a radio frequency pulse

```
%%%%%%%%%%%%%%%%%%%%%%%%%%%%%%%%%%%%%%%%%%%%%%%%%%%%%%%%%%%%%%
%%%%%%%%%%%%%%%%%%%% RF PULSE EXPERIMENT %%%%%%%%%%%%%%%%%%%%

clear all;
close all;
xt=0;
px=0;
pn=0;
pxot=0;
pxit=0;
%%%%%%%%%%%%%%%%%%%%%%%%%%%%%%%%%%%%%%%%
Fs = 20000;                  % Sampling frequency
fc=1000;                     % Fundamental frequency
moduladora=[0 0 1 0 0 0 0 0 0 1 0 0 0 1 0 0 0 0 0 ];    %sequence
[xt bit]=rf_pulse(moduladora,fc)
T = 1/Fs;                    % Sample time
L = length(xt);              % Length of signal
t = (0:L-1)*T;               % Time vector
nfft=length(bit);


figure(1);
plot(t,xt,'k');
title('Useful Signal')
xlabel('Time(s)')
ylabel('Amplitude(u)')
grid on;

xt_fft=fft(xt,nfft)/L;
f=Fs*linspace(0,1,nfft);

figure(2)
plot(f,abs(xt_fft),'k')
xlabel('Frequency(Hz)')
ylabel('Amplitude(u)')
title('Useful Signal Spectrum')
grid on;


stream=RandStream('mrg32k3a');
noise=1.0*randn(stream,1,length(xt))';      % contains exactly the same
    values as noisem
xt_noise=xt+noise';
```

```
figure(3);
plot(t,xt_noise,'k');
title('Noisy Signal')
xlabel('Time(s)')
ylabel('Amplitude(u)')
grid on;

xt_noise_fft=fft(xt_noise,nfft)/L;          %Fourier Transform of the
    noisy signal

figure(4)
plot(f,abs(xt_noise_fft),'k')
xlabel('Frequency(Hz)')
ylabel('Amplitude(u)')
title('Noisy Signal Spectrum')
grid on;


reg_vent=0;
t_total = (0:length(xt_noise)-1)*T;          % Time vector
nfft_total=length(xt_noise);                 % FFT length total vector
L_total=length(xt_noise);
vent_sim=200;
num_iter=length(xt)/vent_sim;
sx_final=zeros(1,length(xt));                % reserve for the final vector
tic
for i=1:num_iter
    xt_noise_v=xt_noise(reg_vent+1:vent_sim+reg_vent);
    t = (0:length(xt_noise_v)-1)*T;          % Time vector
    nfft=length(xt_noise_v);                 % FFT length
    L=length(xt_noise_v);
    C4x_uv=func_cum4uni_vertical(xt_noise_v);    % One dimmensional
        component of fourth order cumulant
    Cum_conv=func_conv(C4x_uv,xt_noise_v);
    Cum_conv=(-1)*Cum_conv;

    sx_final(reg_vent+1:vent_sim+reg_vent)=Cum_conv;

    reg_vent=reg_vent+vent_sim;              % update the vector
end


mod_atn=2*sx_final.*sx_final;
b=firpm(10, [0 0.03 0.1 1], [1 1 0 0]);
sal=filter(b,1,mod_atn);
sal=sqrt(sal);
atn_final=sx_final./sal;

t_cum=toc
```

```
figure(5);
plot(t_total,xt,'b');
hold on
plot(t_total,atn_final,'r');
title('Comparisson in time Useful Signal-Proposed Algorithm')
xlabel('Time(s)')
ylabel('Amplitude(u)')
grid on;
legend('Useful','Output')

atn_final_fft=fft(atn_final,nfft_total)/L_total;

figure(6)
plot(f,abs(xt_fft),'b')
hold on
plot(f,abs(atn_final_fft),'r')
xlabel('Frequency(Hz)')
ylabel('Amplitude(u)')
title('Comparison in frequency  Useful Signal-Proposed Algorithm')
grid on;
legend('Useful','Output')

figure(7)
plot(f,abs(xt_noise_fft),'b')
hold on
plot(f,abs(atn_final_fft),'r')
xlabel('Frequency(Hz)')
ylabel('Amplitude(u)')
title('Comparison in frequency  Noisy Signal-Proposed Algorithm')
grid on;
legend('Noisy','Output')





for i=1:length(xt)
  px=px+((abs(xt(i)))^2/length(xt));
end

for i=1:length(noise)
  pn=pn+((abs(noise(i)))^2/length(noise));
end

SNR=10*log(px/pn)


for i=1:length(atn_final)
  pxot=pxot+((abs(atn_final(i)))^2/length(atn_final)); ;
end
```

```

pno=pxot-px;
SNRO=10*log(px/pno)

CORRE_INICIAL=corrcoef(xt,xt_noise)
CORRE_FINAL=corrcoef(xt,atn_final)
```

www.ingramcontent.com/pod-product-compliance
Ingram Content Group UK Ltd.
Pitfield, Milton Keynes, MK11 3LW, UK
UKHW051848100926
8556IPUK00005B/524